HOW THE CHIROPRACTOR SAVED MY LIFE

My Journey To Wellness and Beyond

By

Deborah Z. Bain, M.D.

www.purelightpublications.org
http://www.purelightpublications.org
Purelight Publications
Dallas, Texas

Published by Purelight Publications, a division of Seaon Ducote Productions
P.O. Box 720193, Dallas, Tx. 75372

Published in paperback and eBook in 2010

Printed in the United States of America

Library of Congress Cataloging-in-Publication Data
Bain, M.D., Deborah Z.
How A Chiropractor Saved My Life
My Journey To Wellness and Beyond

ISBN 978-0-9825988-1-8

Table of Contents

Dedication v

Introduction vii

Preface xi

Chapter 1:   The System Is Broken 1

Chapter 2:   How a Chiropractor Saved My Life 9

Chapter 3:   Enlightened 19

Chapter 4:   No Turning Back 38

Chapter 5:   There is Another Way 41

Chapter 6:   My Tool Box 48

Chapter 7:   A Simple Solution 56

Chapter 8:   Our Future 62

Dedication

This book is dedicated to my husband, Barry, my continual encourager and friend, who has seen me through it all. To my "pink sisters" and others that have had to journey through a difficult medical illness. To my kids, Chris and Courtney, who inspired me to get well. And to Dr. Chalmers, my friend and chiropractor who tirelessly helped me to become whole, and to whom I will be forever grateful. Thank you so much.

INTRODUCTION

My name is Dr. Matt Chalmers. I am a Chiropractic Doctor, certified in Chiropractic Clinical Neurology, with an extensive education in Applied Kinesiology. I also hold a bachelor in Health and Wellness, which I have complimented with over a decade of study in nutrition, eastern medicine, and alternative health practices.

I specialize in helping people that are either tired of taking medication or their medical doctors have run out of options and given up on them. Basically, I pick up the pieces where normal western medicine either has no more answers or, where it has broken the normal human system down and that system can no longer function normally. This is where we find most of our post cancer patients, all of our fibromyalgia/chronic fatigue, multiple sclerosis, and anxiety/depression cases. The system that normally protects the body was neglected and it finally breaks down.

The journey back can take months or years, depending on the severity of the trauma and the length of time that the problems have gone unchecked and untreated. These problems can be further complicated by layers of poor nutrition, stress, lack of proper exercise, parasites, infection, heavy metals, and time. All in all, by the time a person comes to me they are usually in dire straits. This was the condition Dr. Bain was in when I first met her.

The list of problems she was enduring was beyond what the majority of people could fathom. They included infections that ran rampant in almost every organ, tremendous nutritional deficiencies, large infestations of parasites and Candida overgrowth, toxic chemotherapy drugs stored in her normal body tissues, and scars from several surgeries which would have to be remediated before any real nutritional detoxification could be effective. To complicate matters further, she had the majority of her cervical spine and lumbar spine fused thus decreasing the normal neurological control to her body and brain.

Deborah had started some basic nutrition supplements that brought her just so far. She knew she needed to detox, but did not know where to begin. So I got to work. I must say that she was and still is one of the most open minded people I have ever met. Most average adults would not have done all the different things that she had to do to get well, however, she stayed the course even though a lot of what I was having her do was against what normal western medicine claimed is worthwhile or even possible. We began her detoxification slowly; first the colon, then the upper intestines and kidneys, then the stomach, next the gallbladder, and then the liver. Once the liver was clean we were able to allow the body to utilize its own natural detoxification processes. We just kept the aforementioned detox organs clean and the body cleaned the blood and significantly decreased the amount of toxins in the tissues.

After the normal physiological detoxification was complete, which took several months, we started detoxifying the heavy metals and toxic chemotherapy drugs. Applied Kinesiology testing was crucial for this entire process as it is impossible to tell when to start or how hard to cleanse heavy metals without an Applied Kinesiology test. After several months of heavy metal/chemotherapy drugs detoxification she really started to shine. She lost about 12 lbs of toxins! She started to get not just full hair but now it was shining and looked healthy again, just like a shampoo commercial. Her energy levels skyrocketed.

Deborah's whole attitude and outlook on life began to change as she started feeling better. Deborah began to change. When we would talk about stress she stopped mentioning life in a negative, hard-to-get-through fashion. She would talk about all the things she had to do and wanted to do in a lighter tone, as though she was excited about them again. She was transforming into the person she knew that she had always been; fun, energetic, happy, and a joy to be around.

The entire time that we were working on her nutrition we were working on her chiropractic needs as well. The first day I saw her she told me that due to the surgery no other chiropractor wanted to adjust her neck. However, due to the training I had received while working on specific injuries I was able to provide a number of different techniques that were much safer, and thus could get her neurological component moving along. I had been combining adjustments with exercises for about a month when she told me that she was no longer bothered with the pain in her neck and recurrent headaches that had been present since long before her first surgery. Soon after that, she announced that she was actually having days where she would have zero pain from her back; something that she had decided long ago that she would never experience again.

Two or three months into the chiropractic treatment, I made a very large adjustment on her cervical spine and lower skull. She got up off the adjusting table, put her glasses on and told me that her vision was all fuzzy. I must admit that was disconcerting to say the least. However, when I saw her the following week she informed me that she had gone to her normal eye doctor and he had told her that her prescription was now too high and needed to be reduced. He told her I don't know what you did but your eyes seem to be stronger. After hearing the news I performed a very quick neurological eye exam and discovered that her eyes had reset, thus giving her a different vantage point and making her normal vision stronger. I was very happy to see this because this meant that her cervical spine was not hindering the brain and spinal cord as much as we had first thought.

Deborah has gone through a series of detoxifications that have allowed layers of toxins to be peeled back, like layers of an onion exposing more toxins that need to come out. She is now familiar with a lot of things that she needs to do to keep herself healthy and comes in regularly for adjustments and retesting to check her progress. It has been a long road for Dr. Deborah Bain, however, she toughed it out bravely and came out on the other side better than she went in. I was honored to have been such a large part of her journey; however the credit lies with her for sticking out the hard parts and gutting through the uncomfortable and the weird to get to the other side. I can think of no other person I would let treat my children or any other member of my family. I have the utmost respect for not only what Deborah has been through but what she has chosen to do in order to overcome these things. It is no wonder that her patients are so devoted to her. She really does love making lives and families healthy and whole again.

As it was with Deborah and the majority of the patients I see, standard western medicine fails at adequately addressing medical chronic issues, those of over a month or so. These are most often due to a lack of specific nutrients or chemicals your body needs to run normally, or because it is being held back by low grade chronic infections or toxin buildup. These are the things that must be treated in order to return normal function and health to an individual.

I have been researching and learning about nutrition and physiology for nearly 20 years. I have also been fascinated with eastern medicine and the ways that it is possible to look at the body, without thousands of dollars worth of tests, and determine what is wrong with an individual. Applied Kinesiology, or AK, gives me this ability. Through a series of muscle tests that are tied directly to the body's own meridian field, much like in acupuncture but with no needles, it is possible to look at each organ and each organ system and determine what it is that the body wants. Mastering this skill is like mastering a martial art - it takes practice and time. However, unlike a martial art, it also requires people to test and to work with, and I have been very fortunate to work with some very great people such as Dr. Bain.

My only warning for the public is that while supplements are much safer than prescription drugs they are not risk free, nor are all supplements the same. In fact, the vast majority of supplements on the market are not good for you even though they are called "all natural". Make sure that the person you see to give you advice has the proper training and background to give you good safe advice. Look at their track record; have they made many people healthy and do you trust them? Also, remember the vast majority of people cannot detox out of order. Doing any blood, brain, liver, or gallbladder cleanse first, or as your only cleanse, is asking for trouble or at very best asking for zero results. Natural healing can work wonders, however it is important to find someone to help guide you through the junk and the noise so that you can come out on the other side healthy and better than you were before.

Dr. Matt Chalmers, DC

PREFACE

I used to think life was about the destination. But I am now convinced that it is more about the journey and what people we touch along the way. In my journey to recover from nine neck and back surgeries and breast cancer, all in my 30s, I initially found myself lost in the system. I blindly followed the doctor's advice to fuse this and that area of my spine. I was subjected to chemotherapy and radiation. In my medical training, I was taught to believe in a compartmentalized model of medicine. Everything had a diagnosis and treatment or medication assigned to it. But since my treatment, I have found myself questioning the very principals that I lived by in my previous practice of medicine. I turned to a whole network of alternative medicine providers to help me find true wellness, not just Band-Aid fixes for my health problems. For me, it was turning to chiropractors, naturopaths, and nutritionists to get me well. Whatever the path we follow to find wellness, I think we have to be actively searching, not just blindly following. I have gained respect for so many different ways to get someone well. I do not have all the answers to getting someone well as a pediatrician. That is something that I have accepted over the years. I think pride gets in the way of many good doctors, preventing them from being great doctors. If we just all work together and open our minds to many ways of treatment, and focus on the needs of the patient, then we will all be doing what is right. As I have opened my mind, body, and spirit to these paths, I have continued to find wellness in its most comprehensive sense.

As I have journeyed down my path, I have encountered so many people who are on their own paths to becoming well. I feel compelled to reach out to them to show them the way. Not in the same way as I practiced medicine prior to my medical illnesses, but in a very different, holistic way. This book is a way for me to help those who are lost and confused, those who have been wronged by the very system of medicine that was supposed to heal them. This book is part of my ministry of healing. It is meant to open your mind to other possibilities in treating illness. I hope it also serves as a tool to motivate you to become your own health advocate for yourself and your children.

There is another way. Finding the root of disease, or better yet, preventing it from happening at all, is so much more effective than picking up the pieces later and trying to do damage control. It's all about healing disease, not just treating it. Until doctors understand this link and become more integrative in their treatment approach, the patient will be no more than "well managed" with Band-Aids on the real problem that is trying to rear its ugly head.

We all have a journey that makes us who we are, so journey with me as I share some of my own experiences in recovering from a wrecked immune system, as well as how I have used what I learned to help make my patients truly well.

———

Chapter One

The System Is Broken

I had five neck surgeries and four lower back surgeries for herniated discs, all in my 30s. My herniated discs were not a result of any accident, and could not be explained by any major injury. Initially, I developed a severe muscle spasm one morning as I woke up and extended my neck. My neck seized up, and I developed a tingling sensation and weakness in my left hand. I tried physical therapy, but it just made things worse. I was diagnosed with a level C5-6 disc herniation and was told that a "simple fusion" would fix the problem. I went through surgery and recovery and within a few weeks felt as good as new. I went back to business as usual.

Just six months later, I stood up, after filling my cat's food bowl, when I felt a pop in my neck. Pain and progressive weakness rapidly developed down my left arm and hand. Four days later, I had my second spinal fusion, this time C5-6-7. Again, the doctor said a herniated disc needed to be fused and the problem would be fixed. Just two and a half months months later, I was hanging some clothes, above my head, in my closet, and I felt the eerie all-too-familiar feeling in my neck that something was not right. Within minutes my left arm, from my biceps down to my wrist and fingers, was tingling and weak. I could not flex my arm or grasp anything with my left hand. Five days later, I had my third surgery, this time because the doctor said the nerves from the previous fusion were not completely decompressed.

I was able to return to working, seeing patients and taking care of my family. My son was twelve months old at the time of my first surgery. Each surgery took a toll on my health, my family, and my pediatric practice, but I thought that each one was going to be the last and I would be able to get back to my normal life. My thinking back then was so compartmentalized. I think that is why I was so blinded by the recommendations to have surgery instead of seeking other treatments, or better yet, ask the question of why I had the neck of a sixty year old at such an early age.

I tried getting back to life as usual, but I had a constant fear that anything I did would trigger another disc herniation. The ultimate test was a ski trip to Colorado that my husband really wanted us to take. Five months had passed since my last neck surgery. My orthopedic doctor cleared me, and led me to feel that my neck was stable with plates and wires in place. He said I should go and have fun. One of my worst fears was realized when I developed severe neck pain, and weakness in my left hand, biceps, and now triceps while on the trip. I was diagnosed with a new C3-4 disc herniation when we got back. Even the orthopedic surgeon decided to offer conservative treatment this time, with steroids and a seizure medication called Neurontin that is prescribed for neuropathies. It alleviated the problem, and my weakness resolved temporarily. Ten months later, however, I had progressive neck pain, left shoulder weakness, and bicep and tricep fatigue. My weakness was so extreme that I could not even lift my blow dryer above my head. So I was back to surgery for the fourth fusion, this time between C3-4-5. I now had two metal plates holding two large bone masses together, and a wire running along the back of my neck to keep everything in place. It did not dawn on me until much later that the concept of neck fusion does not make a lot of sense. If you fuse one level of your spine, it creates an area of abnormal, increased movement and stress above and below that level. I realized that the pain was not taken away just because the problematic level was fused. Another painful lesson I learned was that if you have two large bone fusions in your neck and you do not do anything about the space in between them, ouch! This lesson was learned all too well one day when I was gardening. I developed tingling and weakness in my left thumb. These symptoms developed one and a half years since my last surgery and just ten months after we managed to have a baby girl, Courtney. I was so in touch with my nervous system by now that any sign of weakness or tingling sent me straight to the orthopedist for more scans. Part of me knew that something was just not right about my last surgery. That was confirmed when more scans revealed a pseudoarthrosis, or non-union, of those two bone masses in my neck. I had grown to tolerate my neck pain as a part of life, always there and debilitating in its own way. I had my final neck fusion, taking both bone plates off and having a tiny plate secured in the space between my C4-5 vertebrae to help fuse the bone that had not grown together. Finally, after three and a half years, my neck was stable and the pain was gone. At least I had no residual weakness after going through so many surgeries, and I thank God for that.

I wish I could say that was the end of it, but I went on to develop lower back pain and recurrent tingling in my left leg. Then, just two and a half months after my definitive neck surgery, I developed a sudden onset of left foot drop and great toe weakness. More scans, more medications, and more surgery were prescribed for what was explained to me as "hypertrophied facet joints" that were pinching off my sciatic nerve. Why was I not asking the hard questions? Why was I so blinded by the system as to believe that what the doctor was telling me was the only way? Well, I have tried to block the seventh surgery from my mind. It was to repair a dural tear and spinal fluid leak after my back surgery. I was in excruciating pain, and could not stand up. I fainted at church because my spinal fluid was leaking out into a pouch under my skin, draining out of my brain. I continued to have persistent back pain, recurrent foot weakness and lateral lower leg weakness. I was constantly informing the orthopedist of my latest symptom. I did this not to have him do more surgery, but hoping he would put all the pieces together as to what was going on with my spine and why it was happing. Instead I was sent to a pain management doctor who provided symptomatic relief of my pain with steroid injections. Six months after the second lower back surgery to repair the spinal leak, the doctor felt a 360 fusion was necessary because of my persistent symptoms and my leg turning blue and cold. A 360 fusion means that at surgery the surgeon fuses the posterior back then flips the patient over to fuse the front through the abdomen. I guess the orthopedic surgeon wanted to make sure that the joint did not move at all by fusing it 360 degrees around. The surgery itself was brutal. I felt literally like I had been cut in half. It was the most painful, longest recovery I had ever to experience, trumping my spinal leak by far.

I have not spoken much about my kids. My son was now five years old and my daughter, twenty-two months. It was hard enough to recover by myself, but with a husband, family, and pediatric practice, there was little downtime. Then six months after the 360 fusion, one of my worst fears was realized. I was caught off guard by my son who playfully pushed me off the couch, loosing my balance I fell and landed on my tail bone, jarring my spine. I experienced progressive weakness all over again, and within two days I could not lift my left leg off the floor. I shuffled around like a little old lady. Despite steroids, Neurontin, and steroid injections, I was headed for a ninth surgery to decompress L3-4 through L4-5.

During all my evolving neck and back problems, I had multiple steroid injections, oral steroids, and anti-inflammatory drugs to ease my chronic pain. In so many ways I can relate to the vulnerable, chronically ill or chemically affected children I care for, because I, too, have been a victim of the broken medical system, putting Band-aid fixes on my health problems and facing toxic overload from the medications prescribed. As far as the "system" was concerned, I was cured after each spinal fusion. But I felt less and less healthy, more fatigued, and beat down with each surgery.

Then, just nine months after my last back surgery, I found two lumps in my breast. I waited two weeks, hoping it would just go away, but it didn't. My Gynecologist sent me for an ultrasound and mammogram, reassuring me that it was probably nothing. I remember the overwhelming fear and panic that came over me as the radiologist said, "I'm sorry, ma'am, but this looks malignant." I remember sitting in a back room crying, waiting for my husband to arrive. The radiologist said he needed to perform a biopsy on the lumps right away. The biopsy felt like harpoons being stabbed into my breasts. Three tumors were found on the right breast, with microcalcifications seen on both sides, so the radiologist performed a biopsy on both sides. On the left, he hit an artery, causing a huge hematoma. I left the diagnostic center changed forever. I had ice packs on both breasts and was in excruciating pain. I came home and had a major panic attack. I felt a lump in my throat, my heart was racing, and I could not catch my breath. For the first time I felt like this was something out of my control. With my neck and back surgeries, at least I understood that there was something that just needed to be fixed and I would be fine afterwards. It was then that I got down on my hands and knees and asked God to give me the peace and strength to get through this, whatever the results may be. Words cannot adequately describe the incredible feeling of reassurance and peace that came over me, knowing that God was in control and that I should not fear.

The biopsy results came back on my thirty-ninth birthday. I had Mammary Cell carcinoma. The radiologist called while I was seeing patients in my pediatric clinic to give me the news. I remember feeling numb all over. I think I was in shock as I drove myself home that afternoon. I had so many feelings running through my head, from anger to panic to confusion. I started trying to find someone or something to blame for my cancer. I read numerous articles pointing to radiation exposure as a cause of cancer, and I remembered that during all my x-rays and CT scans for my neck and back problems, my breasts were never shielded from radiation. Just as I was stewing about the possibility of radiation causing my cancer, I met several women with breast cancer who seemed not to have any risk factors. It was not an accident that I crossed paths with these women. Who was I to be angry? It was then that I began concentrating on getting well, not wasting energy being mad. There would be another time to put it all together and have revealed to me what I may have had in common with these women.

Just ten days after my diagnosis, I had a double mastectomy, and shortly after that I started chemotherapy. As I look back, I can say my cancer was predictable. My immune system was shot; I had all the signs of adrenal exhaustion, and so much tissue damage and radiation exposure from all the procedures that had been done to me. I had warning signs that my immune system was in trouble, such as recurrent sinusitis, bronchitis infections and sugar cravings. But back then, none of the medical doctors I was seeing were looking at the "big picture" to help me get well. They pronounced me cured of cancer after yet another surgery and after the chemotherapy and the radiation were completed.

Deborah Z. Bain, M.D.

I had eight rounds of chemotherapy. The first round started one month after my mastectomy. The session itself was not terrible. I surrounded myself with several friends to make the best of my circumstances. I developed nausea, bone pain, and chills that first night, but I felt better by the time I had to see patients Monday morning. I had some beautiful wigs designed, so my patients were none the wiser when I started losing my hair; it was slow at first, and then it came out by the handfuls. I had initially planned my chemotherapy sessions to be on Fridays so I could be back at the office on Monday for business as usual. Very soon after chemotherapy was started, it was obvious that I would lose all of my hair. When I started wearing my wigs to work, I had so many compliments on how nice my hair looked. I was able to maintain the image of health for just so long. But I was exhausted when I came home at night. I relied on my church friends to provide dinners for my family and help with housework and kids activities, sometimes two to three times per week. As I look back, I was still trying to stay in control in my out-of-control life.

Chemotherapy number two brought its own unique challenges. I developed a bulging vein in my neck, was diagnosed with a blood clot in my subclavian vein and was treated with Heparin injections and oral Coumadin. Then, as if things could not get worse, I started vomiting on Monday night after chemotherapy and had night sweats. By Tuesday morning I was a mess. I went to work anyway, but I was vomiting at the hospital during rounds and at the office between patients. I was crying about how hard it was to work during chemotherapy. One of my sweet families brought a Ty Beanie Baby gift to me that day. It had the saying on it, "God Bless Our Doctor." I left the office at noon, and went straight in to the clinic for IVF after almost fainting due to dehydration. I do not know how I even got to the oncology office that day. I had two liters of IVF before I could sit up without feeling dizzy. One of my good friends, a fellow "pink sister," was with me during this time. I remember her praying over me as I felt like I had hit rock bottom, and she said something profound that has stuck with me ever since that day. She told me, "Only God knows how today will end, and tomorrow will begin." I felt like a new woman after that day, both physically and spiritually.

6

As I was resting at home, I could not help but wonder how I was going to pull off working and undergoing chemotherapy treatments. Then the phone rang. It was the physician liaison for the company I worked for, offering me a short term disability leave of absence. I took it in a heartbeat, excited to rest finally and take care of myself, something that I never really gave myself the luxury to do during my neck and back surgeries. It was during this leave of absence that I not only became closer to God, but I also opened my eyes to ways that I did not understand. A verse of a song comes to mind that helped me through these troubling times:

Who am I to understand your ways? Who am I to give you anything but praise? Who am I to try to solve the mysteries? You are the heart and soul of all that I believe. I am your child…

I got through my first rounds of chemotherapy with Cytoxan and Adriamycin, and then did four rounds of Herceptin and Taxotere. The latter two drugs did not cause the nausea I had experienced with the others. However, Taxotere had the potential for causing peripheral neuropathy. To take this medication, I had to put my hands and feet in buckets of ice for the entire 90 minute treatment, to divert the blood flow from my hands and feet in hopes of preventing this side effect. At the last chemotherapy session I had a pink basin for the nurses to sign that said "Ice Ice Baby."

After my eight rounds of chemotherapy, the oncologist wanted me on Tamoxofen for five years, even though it gave me extreme joint pain. After a year of taking it I just did not believe I should be on it anymore. Now I was really beginning to put things together regarding disease and the body's inherent ability to heal itself. I became more frustrated at the system with its compartmentalized, drug-heavy approach to healing, and decided that I had to be my own health advocate to get me well.

As I received the many interventions to kill the cancer, and suffered through so many side effects of the treatment, I relied on music to keep me going. Song lyrics like, ***"Here I am once again, I am in need of resurrection, only you can take this empty shell and raise it from the dead. …You can take the pieces in your hand and*** *make me whole again…" and "God turns worthless into precious, empty into full, and broken into beautiful."*

I felt broken, beat down, but getting stronger every day through my trials, with a seed planted in me that would grow and flourish into God's plan for the rest of my life. I felt truly called to share what I had learned, even in the midst of my treatments. God was calling me to look beyond myself and beyond my training.

I had to embrace ideas that were foreign to me in my medical training, such as seeking the help of naturopathic and chiropractic doctors to aid in my detoxification and recovery. I also had to learn about nutrition, something I just did not learn much about in medical school. I have often thought how interesting it is that as pediatricians, we do not have more training in preventive medicine and nutrition; after all, poor or inadequate nutrition is the root of so many preventable diseases.

I do not have all the answers, but I do have lots of questions. With that in mind, I started my journey to "wellness and beyond."

***

Trust in the Lord with all your heart and lean not on your own understanding; in all your ways acknowledge him, and he will make your paths straight. - Proverbs 3: 5-6

Chapter Two

How a Chiropractor Saved My Life

The seed was planted during my first round of chemotherapy that I would need more than just medications and medical intervention to kick this cancer. The chemotherapy and radiation killed the cancer cells, but it also left tissue damage, toxins, heavy metals, free radicals, and "garbage" behind. It is interesting to me that doctors can declare you healthy simply when the "cancer is gone" or the surgery has been performed. I know by experience, that that is just not true. I really did not feel "healthy" after any of my neck or back surgeries; quite the contrary, physically, I was probably worse with each one. I did not start on the road to being well until I addressed some very foundational principles of diet, detoxification, and stress reduction.

Prior to the diagnosis of my breast cancer, and all through my 30's during all of my back and neck surgeries, I had a diet that was composed of microwaved TV dinners several times per week, Diet Dr Peppers twice a day, granola bars and fruit cereal bars for breakfast in my car, and maybe an apple a day "to keep the doctor away." I thought I was doing well not buying food at fast food restaurants. I had no clue that my "packaged convenience foods" were just as bad for me. Not everything I used to eat was unhealthy. I did eat salads and a few other vegetables too. This is not how I was brought up either. I think my mom was just better at making us eat a well balanced meal at each sitting, even though some of the foods even she thought were healthy, I have since learned are not. Fast food restaurants were not even a thought for the dinner time meal back then.

Prior to my diagnosis of breast cancer, I started just running out of energy at my previous pediatric practice around three o'clock every day. I started eating M&Ms as a pick-me-up. It was a sort of joke that I had to have them to get through the busy afternoon of patients. I thought I was staying thin and trim because of being a busy pediatrician on the run. I look back in horror as I would drive through the fast food restaurant line for my kids "happy meals" when I would leave the office late, and then put a TV dinner in the microwave for me, and I would declare that dinner. Just before my cancer was discovered, I was craving ice cream with chocolate and caramel on top. I would learn much later, that my sugar cravings were very much related to the candida that was raking havoc on my immune system.

After my second round of chemotherapy, I had hit rock bottom. My strength was gone, my hair was too, and nothing in my diet could sustain my nutrition enough to heal me. I was so nauseated after some of the chemotherapy, I would just choose not to eat. I knew I had to do something. So I called a friend of mine who became a naturopath to solve some of his own chronic health problems. He started my journey to whole food nutrition as foundational in the body healing. I completely gave up the Diet Dr Peppers,TV dinners and highly processed food. I began a whole food fruit and vegetable supplement that enabled me to "eat" the recommended nine to thirteen servings of fruit and vegetables a day. The more I read about nutrition and its relation to disease prevention, the more I wished my doctors would have just asked some simple nutrition questions or maybe just mentioned taking a few supplements to improve my health.

The truth is, numerous articles in medical journals in the past decade have consistently reported the benefits of eating whole foods as a means of preventing a variety of ailments. It is estimated that two-thirds of cancers could be prevented by proper nutrition, i.e., a diet rich in fruits and vegetables. Unfortunately, most of us find it impractical to eat the nine to thirteen servings of raw fruits and vegetables each day that are recommended for optimal health and wellness. By the way, that recommendation has been increased due to our crops being grown in more and more nutrient deficient soil.

I learned that processed foods, low in nutritional value, consumed over long periods of time cause cancer and other diseases by means of free radical damage. In order for us to carry out life's basic processes, we must burn fuel for energy. Burning fuel is called oxidation. The byproducts of oxidation are called free radicals. Much like the smoke from a fire, their production cannot be avoided as long as we are alive and burning fuel. The free radicals lack an electron, and once created, will attack our own bodies, in effect, stealing an electron from healthy tissue to fill their need. Our bodies repair much of the damage at first, but over time, this process takes a toll. This was an amazing light bulb moment for me. The answer to why I became sicker and sicker with each surgery was in part due to the overwhelming oxidative stress, free radical damage, and a very nutrient poor diet.

Fruits and vegetables are rich in antioxidants and, when consumed in adequate amounts, they neutralize free radicals before they can accumulate in the body and cause damage. Fruits and vegetables have thousands of phytonutrients which work synergistically to give the body what it needs. As chemotherapy and other tissue trauma produces oxidation and oxidative stress, if you do not have antioxidants to lap up the free radicals, they produce tissue damage and further damage to your DNA and cells. Interestingly, the disease process depends on which healthy tissue is damaged by free radicals. DNA damage causes mutations leading to cancer, as in my case. Damage to the nerve sheaths results in multiple sclerosis. Damage to the endothelium, the inner lining of blood vessels, causes plaque formation resulting in stroke and heart disease.

I also began a fish oil supplement daily. As I became more well read on nutrition, I learned all the wonderful benefits of essential fatty acids (EFAs). I learned that EPA, Eicosapentaenoic acid, an Omega 3 essential fatty acid, is a more general cellular healer, involved in rebuilding cell membranes, lowering cholesterol, improving eczema and allergies. While DHA, Docosahexaenoic acid, an Omega 3 EFA, is more for brain, memory, and attention. I started taking fish oil around the time I started radiation. I immediately felt better, had more energy, my mental fog, the so-called "chemo-brain," lifted my skin color improved, and I was able to sail through radiation treatments. I remember others dragging themselves to radiation, very sallow looking skin, dull hair, but for 6 weeks of radiation treatments, I bopped in, received the treatment, and said, "see you tomorrow." I was definitely starting to see the light, from the key role of nutrition in my healing, but also my evolving purpose in life to spread the message of the wonderful power of nutrition in healing.

The saying "you are what you eat" struck a chord in me and made me realize that as I was struggling through my health problems, that I was also teaching bad eating habits to my kids because of my lack of planning. All too often prior to my breast cancer diagnosis, I would be rushing home from the office, would stop and pick up the kids with no idea about what we would have for dinner, then I would succumb to the temptation of driving through the enticing Golden Arches for a happy meal. It did make them happy, but what was I teaching my kids? On my leave of absence, we initiated family meals again, meal planning, prayer at the dinner table, and eliminated a lot of the highly processed foods from our whole family's diet. We started my kids and husband on the whole food supplement too. It is not a coincidence, that in doing so, my son's ADHD disappeared. Could it have been his diet of Pop-Tarts in the morning, school lunches of pizza and chicken nuggets with fries and ice cream for dessert, and fast food for dinner that created his poor ability to pay attention in class? In our rushed society, it is so easy to fall into this trap. I know I did.

So to me, it boils down to two foundational principals that we need to heal and thereby become well. The first is "Good Stuff In" and the second is "Bad Stuff Out." Can it be so simple? I have used this model on the most complex medical cases, and I have to say, it keeps me focused on identifying the child's problem and focusing on a holistic model to help them to get well. The "Good Stuff In" is your foundational nutrition, full of essential fatty acids, minerals, vitamins, and lean proteins. It is your fruits, vegetables, nuts, seeds, and unprocessed or minimally processed foods. Let's face it, the average American diet does not even come close to achieving the daily nutritional needs, and in many ways, is actually bad for you. This processed convenience food era that we are in puts more of an emphasis on fast food than it does on true nutritional value. I learned this lesson all too well with my diet of TV dinners, granola bars, and Diet Dr Peppers that were staple items in my diet. More and more we reach for foods that are marketed to us as "fat free" or low in sugar, but yet, most people are not reading labels to find that "fat free" is actually higher in sugar than the regular product, and "sugar free" is actually sweetened with an artificial sweetener that is potentially detrimental to your health.

Getting healthy required a radical change in my diet. I became a label reader. I first started with a basic foundational nutrition of unprocessed or minimally processed foods with lots of raw fruits and vegetables of all different colors. I made a personal choice to stop dairy and processed wheat products. I made fruit smoothies with added flax oil and a rice based protein powder to sustain my energy. I started drinking only purified water, using glass instead of plastic containers to drink from. I went back to boiling water in a kettle instead of microwaving it. As a family, we started cooking from scratch using recipes from healthy cookbooks. My husband even took charge of meal planning a whole month at a time so we don't have to guess what we are having for dinner any more. We stopped eating any instant rice and pasta dishes that were an easy side item to throw in with a piece of meat and vegetable. Now we cook flavorful whole grain rice and make homemade soups and our own pesto and Alfredo sauces. It is actually fun and easy, and just took some getting used to and planning. We make extra and freeze it which makes it convenient and healthier. We also avoid heating foods in plastic containers. I am sure as a family there are more things we could be doing, but we are healthier as a family today than we were just a few years ago, and I am ok with that. My kids know the concept I teach at the office, called Traffic light eating. My daughter, when she was four, would not go on a preschool field trip to Sonic for lunch because she knew it was not a healthy choice for her. At school parties, she would come home and tell me how much junk food was there and one day when I went to one of her parties, I caught her with a plate of baby carrots! Now that is something I can be proud of. I made good nutrition a way of life, not a punishment or a "diet." Once you are eating right, your body just starts craving those good foods. I no longer crave any ice cream or candy like I did before my diagnosis. And my kids just don't ask to go to fast food restaurants. They just know it is something we don't do any more.

That brings me to the concept of the "Bad Stuff Out." It is in the most basic sense, about removing or eliminating toxins which may be ingested, inhaled, or produced by the body. This includes pesticides, industrial byproducts, etc. The world is a very toxic place. In a broader sense, "bad stuff out" may also include a stressful lifestyle, poor sleep habits, built up negative feelings, etc. All these things are toxic to the body in their own way.

About two years after my last radiation treatment, I noticed that what had been working with my nutrition and my fruit/vegetable concentrate and fish oil was just was not enough anymore. I had hit a plateau in my healing. I was tired, my hair and nails did not look that healthy. I had a saliva test to check my adrenals, and confirmed that I still had major issues with my adrenals with low cortisol levels. I started taking adrenal support supplements from the health food store, but just did not know where to go from here. This is where my chiropractor saved my life. He pointed out to me that I had such a toxic burden from oxidative stress, tissue trauma, and chemotherapy drugs that I had to detoxify from them in order to get well. It had never occurred to me before that the body just could not get rid of these toxins. He was an instrumental part of my healing. He introduced me to the concept of AK or applied kinesiology. This testing involves analyzing the body's Quantum Energy Biofield to identify stressed organs and direct specific nutrients to treat them. It also is a method for determining interference fields such as scars, previous areas of trauma, injection sites, and basically anything that disrupts the normal flow of bio-energy in the body. I truly had to open my mind to a completely different way of thinking to grasp this idea. I embarked on a long journey of healing my organs that were so damaged by chemotherapy, radiation, and tissue trauma from over a dozen surgeries. He would test me every one to two weeks initially to figure out what my body needed. I was awed by the test. He could actually test my adrenals, kidneys, liver, gallbladder, sleep points, etc. and direct his supplements to those organ systems. I felt so much more in control of my health by this approach. It made sense to me that if the organs are stressed and not doing their job to pull their weight in detoxing, then the whole body suffers the consequences. The key was to get all the organs healthy, then work on detoxification, which was no small challenge.

I had several mudding procedures at the sites of my injections, scars, radiation exposure area, and past injury sites. I felt like an onion, peeling away the layers of toxins that had affected me. The "mud pack" therapy used moor mud, shilajit, volcanic clays, and other synergists to "permanently re-establish the healthy energetic flow" through the previously traumatized areas of my body. It was a more direct way of bringing detoxifying agents right to the area of local toxin buildup from bioaccumulated chemicals and radiation exposure. Then I would soak my feet in a detoxifying bath to further remove the offending toxins. With each therapy I became more and more in tune with what my body was telling me. I could sense when something did not feel right and could increase my liver, gallbladder, or kidney support to accommodate for my symptoms.

I began relating things such as my transient increase in the number of hairs I was shedding in a day to an iodine deficiency. Starting iodine drops to help give my Thyroid the nutrition it needed, stopped my shedding. I realized that when my liver was happy, my nails grew strong and long, but when it was struggling with the load of toxins it was clearing, my nails would again become brittle and peel. I had recurrent blurred vision in my left eye and different pupil sizes which cleared immediately with an adjustment. Recurrent blurred vision was also noticed when my liver and gallbladder were stressed during a detox.

Several times during my detoxing, I would just stop being able to sleep at night and would wake up three and four times. I came to learn that it was my clogged up gallbladder that was creating the insomnia, as it could not handle all that the liver sent its way. I would increase my gallbladder detox and support, and within 2 days, I was sleeping again. Although these side effects were annoying, they were all part of the process of getting well, shedding years of toxins from my body and helping me become more in touch with the body's awesome healing power. If I relied on my labs from the oncologist to measure my progress, I would be no healthier today because my labs showed that my kidneys and liver were perfect.

Not only did I aggressively clear toxins from my body, but I also underwent food sensitivity testing. I think this is a key part of becoming optimally well. If you are eating all the wrong foods that are causing ongoing inflammation in the gut, absorption suffers. I ended up with twenty-three food sensitivities and an actual allergy to candida. No wonder I was so sick. Every bit of sugar I ate was just feeding the yeast in my gut. Eliminating those foods gave my body a much needed break from the offending pro-inflammatory foods. I felt better during this restricted diet, and was able to reintroduce most of these foods after a six month period. Food sensitivities and allergies are tricky and somewhat controversial, but it makes sense to eliminate the allergens for a period of time, and heal up the gut during that period, so that when you reintroduce foods, you will have a better chance of clearing the allergen.

I am convinced that if you do not keep your gut healthy, you cannot be healthy, for 80% of the immune system is reportedly located in your gut. Not only that, but the gastrointestinal tract is perhaps the most vital component of the body's detoxification systems, according to Klaire laboratories, a nutraceutical company that specializes in hypoallergenic nutritional supplements. They have an integrative approach to GI detoxification that involves enhancing intestinal motility, neutralizing toxins, and restoring microbial balance through prebiotics, probiotics, digestive enzymes, antioxidants, antibiofilm enzymes, and botanicals. First, prebiotics stimulate the growth and metabolic activity of beneficial microorganisms in the GI tract, contribute to the detoxification benefits of probiotic, and accelerate intestinal transit time. Probiotics are the good bacteria in the gut. They bind toxins such as bisphenol A (BPA), aflatoxin, and heavy metals, and facilitate their elimination, and reduce levels of toxin-producing bacteria in the gut. They also support normal intestinal motility and reduce gut ammonia production. Digestive enzymes aid in digestion and promote improved nutrient absorption and improve protein breakdown, thus reducing potential allergen formation. They also promote carbohydrate digestion so as not to feed the intestinal pathogens. Botanicals such as berberine, caprylic acid, green tea, usa urvi, and chlorella, provide extra support in establishing a favorable balance of intestinal microorganisms. They help to rebalance intestinal microflora, bind intestinal toxins, and discourage growth and activity of pathogenic fungal, vial, and bacterial organisms. Lastly, is the concept of biofilm and antibiofilm enzymes. Biofilms are complex protective matrix layers in the gut that protect bacteria from disruption. Antibiofilm enzymes help disrupt the GI biofilm embedding of potential pathogens. In other words, it dissolves the sticky matrix, allows for the elimination of pathogens with the help of antifungal, antibacterial or botanical treatments, then with the help of the prebiotics and probiotics, the more beneficial colonization is restored. Why is this important? Because the abnormal bacteria or candida that overgrow in the gut create metabolic dysfunction and toxins all their own which in turn contributes greatly to the development of many of the chronic diseases we now see even in our children such as asthma, food allergies, eczema, and autism. During all of my neck and back surgeries, I was on and off antibiotics regularly, not to mention the 2-3 courses of antibiotics I took for bronchitis I would catch every winter. Antibiotics are not benign, and the use of them for an acute infection may pave the path for unwanted pathogens to take over in the gut if proper thought is not given to keeping things in balance and prescribing a probiotic at the same time. However, most probiotics are killed by the antibiotic as soon as it is given, despite separating the timing of its administration. Sacromyces boulardii, a beneficial intestinal yeast, however, is one of the only probiotics that is not killed by the antibiotic, and will keep the gut flora stable while taking them.

Not only are some of the foods we eat, the environment in which we live, and the byproducts of an unhealthy gastrointestinal tract toxic to our health, but so is the stressful lifestyle we live. Although small amounts of stress are beneficial to help motivate you to compete a project or meet a deadline, or get you out of troubles' way to avoid an accident, chronic stress exposure is detrimental and can lead to serious health problems. Chronic stress disrupts nearly every system in your body. It can raise blood pressure, suppress the immune system, disrupt the digestive system, increase the risk of heart attack and stroke, contribute to infertility, sleep problems, and autoimmune diseases, cause or exacerbate skin conditions such as eczema, and speed up the aging process. Long-term stress can even rewire the brain, leaving you more vulnerable to anxiety and depression.

When I was on my six months leave of absence during my cancer treatment, I finally rested. Prior to my leave, I was always on the run, did not take care of myself as well as I should have, and put everyone else's health before mine. With each neck surgery, I would return to business as usual in four to six weeks, not fully healed, going back to my same stressful lifestyle, and not changing anything about my nutrition. Luckily, I have learned from my old ways and have embraced a different way since my cancer treatment. I get regular chiropractic adjustments to keep my moving parts in alignment. I exercise regularly to tone the muscles that connect those bones. I recognized that the lifestyle as a busy pediatrician, mom, and wife, is inherently stressful, but I have finally accepted the liberating fact that I will never be able to completely empty my in-basket. Something is always being added to it. I have learned the hard way that it is so much easier to keep the body healthy than it is to pick up the pieces to get well after a major illness. That is why I work so hard at not only keeping me well, but showing families in my integrative practice how to stay well by teaching foundational nutrition and minimizing stressors to their very fragile immune systems. I may not have been able to prevent the devastating diagnosis of breast cancer in me, but I can surely be a part of preventing some very devastating diagnoses like autism in the children I care for.

***

"I will instruct thee and teach thee in the way which thou shalt go: I will guide thee with mine eye." - Psalm 32:8

Chapter Three

Enlightened

All through my breast cancer journey, I felt God was calling me to a higher purpose, and that all my suffering was not in vain. I had to learn how to get well myself, so I could teach this to others. I created Healthy Kids Pediatrics to do just that. I am continually on that journey to improve my health, as my chiropractor can attest to.

I cannot help but be an active advocate for the holistic way of practicing medicine. It only makes sense. Whole food nutrition, decreasing toxic exposure to the environment, minimizing vaccinations, and embracing a nurturing model of medicine is just the right thing to do. I was not brought up knowing the holistic way, but I sure embrace it now. Doctors are born teachers, and we have the power to make a real difference for health, not only for this generation, but for many generations to come. With the grim statistics staring me in the face of all the chronic diseases that are on the rise in our children, I cannot help but want to make a difference and prevent some of my kids from becoming a statistic. Too many doctors have lost their ability or desire to teach wellness, and have become so bogged down in the band-aid fixes, that they have become part of the perpetual sickness model of medicine as we know it.

I have incorporated a different set of ideas, "tools" so to speak, into my "toolbox." I have found that western medicine does not address all the symptoms that my patients present with. It is with open eyes, not blinders on, that I see the problems my patients present with and come up with real solutions that more adequately address their problems, whatever they may be. If you just, for example, treat recurrent ear infections with antibiotics and do not once think that the milk-based formula the child is on is the underlying reason for the ear infection, then that child may have many courses of unnecessary antibiotics and possibly surgery to put in pressure equalizing tubes as a result. The child would have benefited from the doctor having "IgE food and environmental allergy testing and IgG food sensitivity testing" in their toolbox, not just the antibiotic dejure.

In his book *Healing the New Childhood Epidemics,* Dr. Kenneth Bock addresses the four conditions whose exponential growth in the last twenty years is already in epidemic numbers - Autism, ADHD, Asthma and Allergies. According to Dr. Bock this toxicity is the result of not one thing but a combination: genetic predisposition, the increased exposure and assault of toxins on our children's systems, and the deterioration of nutrients in the food supply. He calls these circumstances the perfect storm which has affected the most vulnerable population - our children. In other words, genetics loads the gun and the environment pulls the trigger.

I take care of some of the sickest children in my practice; those that the "system" has basically written off as incurable, giving the patients and families no hope for any recovery. I have seen kids with severe eczema head to toe and have cleared them completely. I have many autistic families that have sought me out because I was not "one of those doctors that said vaccines had nothing to do with their child's development of autism." I have actually had my hand in curing a few through my simplistic approach. I agree with Dr Bock's assessment and others like him, regarding the culmination of symptoms of disease in the over toxic, vulnerable child.

These are the most recent statistics reported by WebMD in 2008 regarding the epidemic of chronic disease in the US:

Percentage of people in the U.S. who have either allergy or asthma symptoms: 20%.

Percentage of the U.S. population that tests positive to one or more allergens: 55%.

Rank of allergies among other leading chronic diseases in the U.S.: 5th.

Odds that a child with one allergic parent will develop allergies: 33%.

Odds that a child with two allergic parents will develop allergies: 70%.

Food allergies have increased by 400% in the last twenty years.

Asthma affects three million people world-wide and has increased 300% in the last twenty years.

ADHD affects 1.6 million American children, accounting for 8% of school aged children, and has increased by 400% in the last twenty years.

Autism increased from 1 in 10,000 in the 1980s to 1 in 91 children in 2009. Generation Rescue did a survey of 9,000 boys in Oregon and California and found a 155% increased risk of neurological disorders such as ADHD and autism in the vaccinated vs. non-vaccinated group.

I am concerned that what is happening with the huge increases in autism, sensory integration disorder, asthma, allergies, chronic otitis media, eczema, and other diseases involving an impaired immune system is that kids are being exposed to too many toxins, too soon, and they cannot detoxify from them, leading to metabolic dysfunction and disruption of normal physiologic functions in the body. I cannot help but relate to these children with what I believe are "wrecked immune systems," as I reflect on my own chronic health issues. As reflected in these statistics, this is the real epidemic. We have traded the vaccine preventable diseases of the past for the chronic diseases of the present.

Exposures to toxins affect our children starting right from conception. More than ever, our kids are growing up in a toxic world from environmental exposures to pesticides, industrial pollutants, plastics, highly processed food diets, artificial sweeteners and flavorings, and a very aggressive vaccination schedule.

The Environmental Working Group, a non-profit organization, conducted a study in 2005 to assess the "body burden" of toxins in newborns. They tested umbilical cord blood from ten randomly selected newborns and found an average of 200 industrial chemicals and pollutants that had crossed the placenta. They included such substances as mercury, pesticides, perfluorinated chemicals (PFC's) which are the active breakdown products of Teflon, Scotchgard and flame retardants. All of these are linked to birth defects, developmental delays, and even cancer.

Plastics are a way of life now. You use them to store your food in, serve your food in, use to feed your baby, and the list goes on and on. But some plastics are harmful, especially when heated. You may have heard more recently about the toxicity of "BPA" and its use in baby bottles and about the leaching of chemicals into the water bottles you drink from and allow to sit in your car.

Recently there has been a push to use more aluminum, glass, or BPA-free containers. Not only do we have a toxic burden from the things in the environment, but we also have toxic exposures in the food we eat.

Dr. Russell Blalock wrote in his book, Excitotoxins: The Taste That Kills, *"It is my opinion, after reviewing an enormous amount of medical research literature, that MSG, aspartame, and other excitotoxin dietary additives pose an enormous hazard to our health and to the development and normal functioning of the brain."*

So many parents do not pay much attention to labels and do not realize the chemicals in the foods they are offering their children. We are often deceived or lured into thinking a product is good for us by the low fat, low sugar, or sugar-free label. But do you realize how little research was actually done on the sugar substitutes before they became such a large part of the food industry. Not only do these sweeteners stimulate cravings, increase your appetite for sweets and other carbohydrates, but they also have real safety concerns. Aspartame has been known to cause headaches, ear buzzing, dizziness, nausea, GI disturbances, weakness, vertigo, memory lapses, behavioral disturbances, blurring of vision. Because Aspartame is found in so many products, it is very easy to overdose without realizing it.

The oldest of the chemical sweeteners, Saccharin, has been clearly shown to promote cancer in laboratory animals, and has since been abandoned by food manufacturers as a sugar substitute.

Sucralose is the newest chemical sweetener, approved in 1998. Although it is made from real sugar, its chemical structure has been altered with the addition of chlorine, making it more similar to a chlorinated pesticide than something we would be eating and drinking. My concern is that we are now seeing Splenda added to juices, chewing gum and even medications! Just like Aspartame, it is permeating our food choices, and there are inadequate numbers of studies to show its short term or long term safety. I have read numerous reports of individuals that have been adversely affected by this sweetener, but there is no ongoing surveillance of this sugar substitute to compile adverse event data.

In my opinion, it is best to avoid all chemical sweeteners and use natural sweeteners such as Stevia, organic maple syrup/sorghum syrup, raw honey, or Agave' nectar, instead.

Not only do chemical sweeteners pose a potential health problem, but add to it other toxins we unknowingly feed our bodies such as Hydrogenated fats, MSG, Nitrites, and Artificial colors/flavors, genetically modified or antibiotic/hormone laden meats, and high fructose corn syrup, and you have a recipe for trouble.

These are just some of the chemicals we are exposed to, and it seems there is no end in sight. More than ever, we need to turn to whole food nutrition, free of artificial ingredients and hydrogenated fats, and get back to basics. We need to look after our own health, read labels, and be familiar with what you are truly eating and drinking.

Even the common medications that have been used for years to treat "symptoms" of a viral illness have been brought into question and have been deemed unsafe and ineffective in children under the age of six years. So where is a parent to turn when their child gets ill? Are they supposed to just ride it out with chicken soup and nasal saline drops? The latest voluntary recall of Tylenol and Motrin have left parents even more concerned about the safety of the medications they give their children.

Although antibiotics kill the pathogenic bacteria causing various infections, they also kill the good bacteria that inhabit the intestinal tract and keep it in balance. It is in this way that the immune system is potentially disrupted and becomes susceptible to other pathogens such as candida and clostridium. So although the acute infection is treated, the stage is set for overgrowth of abnormal inhabitants in the gut which have the potential for causing their own chronic diseases. Yeast deserves special mention here. Not only does it cause thrush and diaper rashes after antibiotic therapy, but it also has the potential to do much more damage by weakening your immune system and through toxins they produce. Recurrent infections develop, each being treated by another round of antibotics, encouraging further yeast overgrowth, and the vicious cycle continues. A diet rich in sugar and other simple carbohydrates just feeds the yeast. "The Yeast Connection" by William C. Cook, M.D., describes the link between yeast overgrowth in the digestive tract and the symptoms of fatigue, headache, depression and many more chronic complaints. Yeast can also potentially mutate human DNA in the intestinal tract. This was interesting given my breast cancer, because during my search to get well, I tested positive for immediate IgE allergy responses to four intestinal yeasts. The simplest way of thinking of yeast, or fungi, is that they can overgrow and mutate cells which in turn can create tumors, or abnormal collections of cells. I have subsequently learned that Candida has many other detrimental effects such as its role in autism, in eczema flares, and behavioral problems. In my practice, I have seen first-hand these effects and have had to open my mind to the reality of them, both personally and in the treatment of my patients.

Several other potential pathogens inhabit the gastrointestinal tract of our youth such as Clostridia and streptococcus. Each has its own unique concerns. Not only are kids exposed to more toxins than ever before, but I am concerned that their ability to detox from them is impaired. Everybody has a certain genetically predetermined ability to detoxify their body. Some people are born with a "big trash can" capacity to detox from all the environmental toxins they are exposed to. Others are born with "small trash cans." Not only that, but some kids have genetically more efficient detoxing pathways than others, better "trash trucks," if you will.

So much research has been done to prove that autism is not caused by the MMR, and other individual vaccines, but researchers are missing this important point of cumulative toxins and impaired detoxification ability in the presence of highly allergic children with family histories of autoimmune diseases and allergies. I see it every day in my practice. I did not have to look very hard to come up with some examples to share.

***

Jake is a now 4 year old twin who came to me with a history of chronic, recurrent otitis media, developmental delays including speech and motor delays, and had been diagnosed with mild autism or pervasive developmental delay. As I looked further into his history I saw the all too familiar signs of a toxic child with an immune system in trouble. He was a 5 lb premature twin with a history of gastroesophageal reflux, formula intolerance, and constipation. He had numerous sick visits, including cold after cold, leading to ear infections and antibiotics, which eventually lead to placement of pressure equalizing tubes. Jake had recurrent severe diaper rashes that would itch and bleed. Mom would take him to the pediatrician sometimes two times in a week for the rash, but the story was always the same. He would be diagnosed with diaper dermatitis and mom would be told to treat topically with diaper cream. He even had a special compounded cream prescribed by the doctors to apply to his rash. He also had bouts of severe abdominal pain, crying fits, and would curl up against the wall grabbing his bottom or his tummy. The doctors would tell mom that he was just fine because his symptoms would resolve after having a bowel movement. In reality, the recurrent rash and abdominal pain was caused by a massive yeast overgrowth in his digestive tract from the multiple courses of antibiotics he had been on.

Jake would also spike high fevers with his shots, so the doctor's office would pre-dose him with Tylenol before they gave them. He would also sleep for several hours and be very fussy for several days afterwards. He had his Varicella, MMR, IPV and Influenza vaccine at his one year checkup. It was after this set of vaccines that he started to decline. He lost interest in all food except milk and goldfish crackers. He would scream for hours until mom gave him some. He would sometimes curl up against the wall and kick and scream for hours and would also chew on the wooden plantation shutters and would trace the wall or lines along the brick with his nose. He had an upper respiratory tract infection diagnosed on the same day as he received the second influenza vaccine when he was 13 months old, then at 14 months he was sick almost continually for a month with pharyngitis and croup, and even had a course of oral steroids for it, but then still received his DTaP, Hib, and Prevnar vaccines at the 15 mo checkup. He developed a fever, hives on his arms and legs causing them to swell and turn purple within one week of these shots. The urgent care clinic just said he had a virus and denied the possibility of an allergic reaction to the vaccines. Jake was referred to Early Childhood Intervention (ECI) at 18 months for not talking, not walking, having eating issues and tantrums despite having PETs placed. All along, the doctors felt his language and motor delays must be from his ear infections. He received a total of 23 vaccines by 18 months of age.

When I saw Jake, I had IgG food testing done which showed a severe allergy to cow's milk, eggs, and wheat. I recommended he start a casein-free, gluten free diet, start Omega 3 EFA, probiotics, and a fruit and vegetable supplement to provide him with foundational nutrition with which to heal. Mom started him on a whole food, minimally processed diet, free of processed sugars and artificial ingredients as well. We treated his recurrent yeast with three months of antifungal therapy.

His recovery was remarkable. He showed a dramatic response to the food elimination diet to those foods he showed allergies. His behavior improved very rapidly, first with becoming more engaged and making better eye contact, then more gradual improvements in his social skills. He has not had any further ear infections or needed to come in for sick visits. He recently had his reevaluation with the developmental pediatrician who removed the autism, pervasive developmental delay diagnosis, but was still not willing to admit that his change in diet and supplements had anything to do with his recovery.

It is important to note here that the common practice of pre-dosing Tylenol (Acetaminophen) for vaccines should not be done routinely as it interferes with the body's ability to make Glutathione, an important element in the detoxification pathways.

***

Another very important story to tell is that of Billy who is now a three yr old child with a history of eczema, seborrheic dermatitis, and gastroesophageal reflux. He also had a significant history of high fevers and lethargy for several days experienced with every set of vaccines he received. He received a total of sixteen vaccinations in the first nine months prior to transferring to our office. Despite his high fevers, mom still believed that she did not have a choice regarding the vaccines and that he was supposed to get them on schedule. I started seeing Billy at one year of age, and the MMR, Pneumococcal, Hepatitis A, and Varicella vaccines were routinely recommended at that visit. We only gave two of the four vaccines which included the Prevnar (Pneumococcal) and MMR. He developed a high fever of 105 after his shots and became extremely irritable. He had been smiling and taking steps towards me at his well check up, but what I saw within one week of his shots frightened me. He had a blank stare on his face, was slumped over at the foot of mom's chair, would just cry or nurse continually and would not eat. He would not attempt to stand, walk, or make any eye contact. Recognizing that he had a severe reaction to the MMR and/or Prevnar, I sent him to an alternative medicine colleague who confirmed the reaction with AK (Applied Kineseology) and treated him with several homeopathic drops to detoxify him from the vaccines. I firmly believe that he recovered from that reaction because of that intervention. He is a healthy three year old now, has rarely been sick since stopping his vaccinations. He has had some residual sensory issues but is otherwise developing normally. He had subsequent tests that showed moderate vitamin deficiencies and metabolic and mitochondrial dysfunction that we are addressing with supplements. He has not been further vaccinated. He has since had a sister who is on a modified vaccination schedule, doing one at a time. She had a reaction to the first Hemophilus influenza vaccine with a high fever, and is no longer receiving that one, but is otherwise tolerating the rest of the vaccines on her modified vaccination schedule.

Vaccines are not benign. All I have to do is look at the endless list of potential side effects and warnings listed on the vaccine information forms we give to the parents whose children are receiving vaccinations. Side effects listed may be as mild as fever, swelling at the site and fussiness or as severe as Gillian Bare', seizures, bowel obstruction, and even death. The vaccine information forms also warn against vaccinating with that vaccine if you are allergic to any component. The majority of people would not know they have an egg allergy, for example, unless the doctor is willing to check for it. Vaccines contain substances such as formaldehyde, aluminum, mercury, phenol, neomycin, and acetone. Although the chemical makeup of vaccines is controversial at best, what is certain to me, is that some kids are just not healthy enough to receive these vaccines on the aggressive schedule that is currently recommended. I am looking out for the little guy, the one who is vulnerable to all these toxins. I am looking out for the child who has a weakened immune system for whatever reason. In my opinion, you cannot put the term, robust 9 pound at birth, breast fed baby on the same vaccine schedule as a 4 pound preterm bottle fed baby with reflux and formula intolerance. And yet, that is exactly what is happening. I believe in custom vaccination schedules for all babies. I encourage parents to have real informed consent by reading about the vaccinations in books like "The Vaccine Book" by Dr Sears or "Childhood Vaccinations" by Dr Feder, prior to the two months check up.

I also do not believe the Hepatitis B vaccination should be done routinely to all babies in the newborn nursery, as the babies have immature livers and often become jaundiced, and are not at risk for something that is a sexually transmitted disease or one that affects people with high risk lifestyle. These vaccinations should be started at the discretion of their pediatrician at follow-up. At the same time, I find it disturbing that premature babies that turn two months of age prior to being discharged from the neonatal intensive care unit, who are barely up to newborn size at discharge, are receiving a full set of two month vaccines just before they are released.

***

That brings up another patient of mine, James. He is a four month old previous thirty-three week premature baby with moderate Gastroesophageal reflux being treated with Prevacid. He received his first DTaP vaccine at the four month checkup, purposely waiting until he reached the gestational age of "two months" before starting vaccinations. Within a week of the vaccine, mom reported a decrease in bottles, not being interested in feeding, and possible wheezing. I did not link these symptoms with a potential vaccine reaction at that time. He had the Prevnar vaccine two weeks later on the modified vaccination schedule we had agreed on. Within three days, James had worsening of his poor feeding and within two weeks, he lost all desire to suck on a bottle or pacifier. He had to be syringe fed. He would act like he was hungry, take a few sips of a bottle, then refuse the bottle and just act fussy and disinterested. Mom took him to see a chiropractor who tested him with (AK) applied Kineseology and confirmed a vaccine reaction. I sent him for electrodermal screening with a holistic provider who, through different testing, also confirmed a vaccination reaction. He was put on homeopathic drops to detoxify from the vaccination, and he returned to normal within two weeks. Even the modified vaccination schedule proved to be too much for this ex-premature infant. I shudder to think how he would have reacted to the standard scheduled vaccines which would have included DTaP, Hemophilus influenza, Hepatitis B, Inactivated polio, Rotovirus, and Pneumococcal vaccines. Mom has chosen not to further vaccinate at this time. He continues at this time to grow and develop normally and has not shown any further milestone regressions.

Let me elaborate on the importance of taking a thorough family history. As we know, some kids have a larger ability to detoxify themselves than others, and this is genetically inherited. But in a busy pediatric practice, taking a comprehensive family history may not be top priority. What we have found though, is that children with family histories of Autoimmune disease or Mental Illness, tend to be poor "methylators" or in other words, have trouble detoxing their bodies. I encountered this recently when I had a twenty month old new patient in for a checkup. He had a dad who was bipolar and a mom with a family history of diabetes and bipolar disorder. They had one child with Autism, one child with speech delay, and the child I was seeing definitely had speech delay and quite honestly, I thought he was bipolar as well by his severe outbursts in my exam room and his atypical aggressive behavior. He happened not to be vaccinated because the parents felt the others were affected by them.

I believe that any baby who has a family history of significant mental illness or autoimmune disease should at least be considered for a modified or delayed vaccination schedule. It is as simple as that. The child is not ready or healthy enough to receive all the vaccines on the current schedule. I know vaccinations have been touted as important advances in the health of our nation, but having eight vaccines in the first two years of life given in the 1980's and to quadruple to thirty-two in 2009 is a little aggressive for anyone. Phasing in vaccines such as the Hepatitis B later in the schedule makes a lot of sense to me and my patients, especially since a recent 2005 study showed that 15 years after neonatal immunization, a large proportion of children exhibited waning immunity. In other words, when teenagers need protection they do not have it.

Dr Lauren Feder, a holistic pediatrician in California adds, "for children who have chronic medical conditions such as allergies, eczema, or recurrent ear infections, the timing of vaccinations becomes more complex. Unfortunately, so many children live with chronic problems now that we also need to consider the possibility that the condition has been caused by or is connected to previous shots." I agree with Dr Feder that it is best to treat these underlying conditions, preferably with natural holistic medicine, before vaccinating further.

Another observation I have made in my practice, is that those families who choose not to vaccinate or do a selective schedule stay far healthier and are rarely in for sick visits compared to the fully vaccinated group. Now granted, these are usually the same families who believe in a whole food diet, use supplements and homeopathic remedies. I think they are on to something. Getting back to the basics and minimizing the chemicals that enter our bodies just makes a lot of sense.

I would like to share with you the story about Sam. He is now a two year old that I have seen since just a few weeks old. He received his Hepatitis B vaccination at one day old, but has been on a modified vaccination schedule at our office ever since. Sam developed moderate eczema while on Nestle Goodstart formula and was changed to Nutramagen. He also had severe recurrent wheezing in the first six months, of his life. He was hospitalized at seven months for severe wheezing and was seen in the Emergency Room at 9 months for the same thing. He has been treated recurrently with Pulmicort, an inhaled steroid, and Albuterol, a bronchodilator, for the symptoms. He was also being managed by a pulmonologist who saw him in the hospital. At the pulmonologist office, he was given a flu vaccine at fourteen months (containing Thimerisol) to which he had an allergic reaction. He developed more intense wheezing, rash, and worsening eczema. We did allergy testing that did not show any food allergies. Sam received a total of 3 DTaP, 2 Hemophilus influenza, 2 Prevnar, 1 Flu vaccine, and 1 Hepatitis B (at birth) for a total of 9 vaccines in the first 14 months of life. During that time he remained sick, on nebulizer treatments almost daily, had recurrent viral illnesses, and his eczema persisted despite negative allergy testing and a hypoallergenic formula. Since deciding not to further vaccinate for the last several months, his eczema has cleared up, he has been off nebulizer treatments, and has had no further Emergency Room visits. Is it a coincidence that allowing Sam to detoxify from all the medications and vaccines was what his body needed all along? Is it not possible that he was really having allergic reactions to the chemicals in the vaccines themselves, as he did with the Flu vaccination?

One theory that explains the chronic medical problems and recurrent infections our vaccinated kids face is the TH1-TH2 imbalance. TH2 is responsible for making antibodies for the vaccinations that we receive. To ensure a vigorous and robust antibody response to vaccinations, an adjuvant is added such as Aluminum. Aluminum has been found to increase the production of IgE antibodies. The body makes tons of antibodies and can cause IgEs to go crazy and cause IgE allergic reactions. The hyper TH2 response may be contributing to why so many vaccinated children are developing allergies, asthma, eczema, and a whole lot of other medical problems. Meanwhile, with the weakened TH1 cells, they are not able to fight off any childhood illnesses and often end up relying on multiple antibiotics to treat recurrent sinus and ear infections, further weakening their immune system. Not only that, but the stimulated immune system is also responsible for the development of food and environmental allergies. IgE reactions are the

most recognized and are immediate sensitivity reactions. IgG reactions
are less well known and refer to reactions that are more delayed, causing
a constant inflammatory response in the body. IgG is a protein made in
the body that normally identifies bacteria and viruses to protect you, but
when it starts tagging food, you have a major problem. The constant
inflammatory response going on in the body by continually eating those
foods you are allergic to has been linked to the brain fog and disconnect
our children, and adults for that matter, experience. Children at risk for
the "leaky gut" that contributes to the development of these food
allergies in the first place are the premature babies, children on recurrent
antibiotics, corticosteroids, nonsteroidal anti-inflammatory drugs, or
children who have limited variety or a diet high in sugar, or those with
gastroenteritis or malabsorption syndromes. A natural progression then
of the leaky gut and malasorption caused by the intestinal inflammation
and food sensitivities, is nutrient depletion, especially of B-complex
vitamins and minerals such as Zinc and Magnesium. Not surprisingly, it
is these same deficiencies that I commonly see in my patients with
autism, eczema, and ADHD.

*To further illustrate this point, let me tell you about Aiden. He
came to see me when he was 2 ½ yrs old. He had been diagnosed with
developmental speech delays and was seeing therapists at a local
Early Childhood Intervention program. He was also showing some
awkward behaviors including repetitive clapping, cramming food in
his mouth, cravings for cheese and carbohydrates, head banging, long
temper tantrums, and repeating phrases over and over or scripting
television shows.*

*I reviewed his all too predictable history. He received his Hepatitis
B vaccine at two weeks, then at two months received his DTaP,
Inactivated polio, pneumococcal, Hemophilus influenza, and 2nd
Hepatitis B vaccine. He developed bronchiolitis at six weeks of age,
and two weeks later had his first ear infection. By the four month
checkup, he had mild eczema. He received his 2nd DTaP, Inactivated
polio, pneumococcal and Hemophilus influenza vaccines. Two weeks
later he had a URI. At the six month checkup, he received round three
of vaccines. Then two weeks later he had another cold which*

progressed to another ear infection. He developed pharyngitis at nine months, then had his checkup and Hepatitis B shot two weeks later. He was diagnosed with an allergic rash at ten months of age. Then, at his one year checkup, he received his MMR, Varicella, Pneumococcal, and Hepatitis A vaccine, and had a TB test. Three weeks later he had another ear infection, received another round of antibiotics. He had a sinus infection at fourteen months, Respiratory Syncitial Virus bronchiolitis, and ear infection at fifteen months, and another ear infection at sixteen months. He had his eighteen month checkup with more vaccinations of his Hepatitis A and DTaP even though he had an ear infection at the time which required more antibiotics. It was not a surprise to me to see noted in his chart language delay and developmental delay at that visit. He had several more ear infections and sinus infections, and was diagnosed with nonspecific allergies around his two year birthday. He had Pressure equalizing tubes placed in his ears at two years.

This case represents so many of the patients that seek me out. They have been affected by vaccinations, but no one is willing to admit it. This is one of those vulnerable children that could have avoided long battles with illness if only someone would have put all the pieces together earlier.

We first did IgG allergy testing which showed level four allergies to cow's milk, level three to wheat, citrus, and cheese, level two allergies to almond, oat, and peanuts, along with multiple others. Just changing his diet and putting him on some key supplements seemed to improve his language and overall behavior and lifted the "fog" he was under.

He has been found to have multiple nutritional deficiencies as well as an IgG deficiency and is seeing a Defeat Autism Now doctor to help detoxify him from heavy metals. He continues to show good progress.

Aidan was a victim of a very broken medical system that deals more with "Band-aid fixes" and having all those vaccinations administered on schedule, than it does on identifying an underlying problem and actually admitting that those things we do to a child to keep them healthy, may indeed be the reason for their illnesses in the first place.

This next story really sums up why I wrote this book and why I voluntarily exposed myself to the scrutiny and skepticism of the medical community. I have learned that the truth can be uncomfortable, but admitting the truth is the first step in affecting real change. I believe that I endured all the procedures, surgeries and ultimately breast cancer and was then made truly well and was enlightened to a holistic way, so that I could teach this message the rest of my days to all who will listen. I believe the rate of chronic disease in this country will never decrease unless doctors embrace a different mindset and way of looking at disease, as a continuum of breakdown of the body's defenses. So here is Lizzie and Noah's story.

***

Lizzie is a now seven yr old recovering autistic child. She presented to my office at the age of four with a history significant for being born by cessarian section at 37 weeks after mom was treated with Terbutaline for2 months for preterm labor starting at 29 weeks. Family history was significant for autoimmune disease, food sensitivities, inflammatory bowel disease, Rheumatoid arthritis, and Alzheimer's and Dementia. The Hepatitis B vaccine was given at two weeks of age. Two weeks later she had thrush and was treated with Nystatin. She developed a green nasal discharge and was put on Amoxicillin one week later. The sores in her mouth worsened and she developed a cough, so she was again placed on Nystatin. Two days later she was prescribed Omnicef and Orapred for the cough. At the two month checkup, she was found to be colicy and fussy and was trying different formulas. She received her DTaP, Hib, IPV, and Prevnar vaccines at this visit. At five months she was smiling at everyone, laughing out loud. She developed constipation and had recurrent abdominal pain. At six months she was noted to have some eczema for which Aquaphor lubricant was recommended. She was given set #2 of DTaP, Hib, IPV, and Prevnar vaccines. Lizzie started flapping her arms crazily when she heard loud noises and would flip pages of books for hours. Eczema and constipation reportedly worsened. At her nine month checkup she was given IPV, Prevnar, and flu vaccines. She started bouncing up and down on her bottom, ran into walls, crawled with her head down. All these things the parents thought were hilarious, but they did not put together that Lizzie was becoming less and less engaged with her surroundings, and that these were self-stimulating behaviors. Just before her one year birthday Lizzie was given a flu vaccine. At her birthday, she screamed and cried and wanted to be in another room. At her one year checkup, she was given the MMR and Varicella. She began vomiting on and off for several days after the vaccines. Then two weeks later was back on antibiotics for an upper respiratory infection. Two weeks after that she was in the emergency room for 105 degree fever and mouth sores. At her 15 month checkup, she was noted to be speech delayed, only saying three to four words. She also never pointed or waved or played peek-a-boo as an infant. She started having repetitive actions such as flapping and talking to her hands. She was given her DTaP and Hib at this visit. Two weeks later, she was back on antibiotics for another upper respiratory tract infection and bad cough. She was also given Nasonex and Rynatan for suspected environmental allergies. Mom also reported excessive crying, fussiness and sweating at naps, and poor eating for 4 weeks after her vaccines. Lizzie continued to have recurrent upper respiratory infections and ear infections and was put on several more courses of antibiotics through her toddler years. At her 18 month checkup Lizzie had lost all functional language, engaged in repetitive behavior and jibberish. She had no desire to interact with people, not even family. At her 2 year checkup, she had outbursts, body tightening all over and screaming, self-injurious behaviors, and severe tantrums. She was referred to a developmental specialist who confirmed the diagnosis of moderate autism.

Mom halted vaccinations and had allergy testing done. Lizzie was taken off the allergic foods and never had another ear infection and rarely had respiratory symptoms or cough develop which had been the norm in her first two years of life. She had nutritional testing which showed severe deficiencies. She also showed high levels of heavy metals and viruses in her body. She started specific nutritional supplements to correct her vitamin and mineral deficiencies and had intensive Occupational, Speech, and Behavioral therapy. As I have taken care of Lizzie for the last three years, she has really begun to shine. She went from having no functional language and having outbursts and hand flapping to being mainstreamed in school with minimal help, speaking in full sentences, is social with many friends, and even sang a solo at church in front of a huge congregation.

Lizzie's younger brother, Noah also showed some serious signs of an immune system in trouble during infancy. He had a history of severe eczema head to toe to the point of open sores. He was a 39 week term breast fed baby born by VBAC (vaginal birth after C-section). He had poor sleep patterns, constipation, and continual fussiness during infancy. He received five courses of Amoxicillin for his inflamed eczema and possible impetigo during the first year of life prior to transferring to my care. At one point, mom described him grabbing and clawing at his neck which she suspected was related to a food reaction. Mom was concerned about food allergies, but no test was offered, and instead she was told to supplement with soy formula and for her to avoid strawberries, peanuts, dairy, tomato, and seafood while nursing. She was also given booklets on the pathophysiology and management of eczema and was treated with Aquaphor, Benadryl, Zyrtec, and topical hydrocortisone.

When I first saw Noah, he was extremely irritable with head to toe eczema that was open, rashy, and weepy with cracking in all his skin folds. I used the "good stuff in, bad stuff out" approach to guide us in his treatment. I ordered IgE and IgG food allergy testing which showed over 25 allergies including milk, rice, soy, egg, and several nuts, wheat, oats, and corn. He did a food elimination and rotation diet which helped some, but I felt that it was an uphill battle because of the sheer number of foods he was reactive to. Dr Chalmers tested Noah and found that he did receive the Hepatitis B vaccine at birth, which he may have been reacting to and did a mudding procedure on him to detox any heavy metals. Amazingly, his eczema cleared up dramatically after that procedure to reduce his toxic burden. Next I ordered a comprehensive stool analysis and nutritional evaluation on him. He showed poor absorption, bacterial dysbiosis (imbalance in good and bad bacteria in the gut), and yeast overgrowth on the stool analysis. His nutrition testing showed severe deficiencies in B-

complex vitamins, antioxidant and mineral deficiencies. He had mitochondrial dysfunction, detoxification impairment, oxidative stress and impaired methylation. In other words, his immune system was a wreck. We have started him on mineral and vitamin supplements to repair his deficiencies and have treated his yeast overgrowth. We are very hopeful that our persistence and focus on healing of his immune system will bring him lasting healing and wellness. I fully believe that if it was not for mom's proactive refusal of vaccinations, in lieu of his metabolic disruptions that Noah would be autistic today, like his sister was. This case is an example of the predictable pattern these kids have that go on to develop autism spectrum disorders. It is in this light that I hope physicians rethink their stance on vaccines and become less rigid on delivering them "on schedule," but rather, focus on delivering a healthy child on the other end instead.

The hope in all these stories I have presented, is that early recognition of a problem shows great promise for reversing. Numerous patients have benefited from specialized testing to determine food sensitivities, allergies, and nutritional deficiencies. If we as physicians do not expand our tool boxes to incorporate some of these newer tests, then we are going to remain blind to what we need to do to truly help our patients.

***

"I can do all things through Christ, who strengthens me." - Philippians 4:13
"On Christ's solid rock I stand, all other ground is sinking sand" - Deuteronomy 31:8

Chapter Four

No Turning Back

If I said my story ended with me beating cancer and feeling closer to God, then I would not be telling you about God's ultimate victory through my journey. God's ultimate will for my life was not for me to become healthy and live happily ever after. No, His ultimate will was realized through the growing passion born in me during my breast cancer journey to share what I have learned about health and wellness with others. I had to go through my physical illnesses to learn how to be healthy myself, to learn balance in my own life (something that I continue to work on), to learn how to "Let Go and Let God," and only then, I would be ready to carry out His ultimate will for my life. I can relate to so many of my overworked, stressed out moms who continue to struggle with balance. I can relate to those same moms driving through the fast food line to pick up "dinner" for their kids, because I was there too. But I can also relate to those kids with autism, eczema, ADHD, adrenal exhaustion, Crohn's disease, and others, because, often, they have been just as affected by the "system" that was supposed to make them well, but failed. My heart aches for these kids. I cringe when I hear some of the stories my parents share with me about how many drugs and procedures their children have had to endure, all for nothing because they are no closer to being well than when they started. I see these kids in a different light, as having usually some underlying metabolic or immune system problem or micronutrient deficiency that can and should be addressed. I do not see them as a "diagnosis" for which a drug is used to patch the problem. With this new wisdom, I could no longer practice medicine in the same way.

I began having vivid dreams about creating an integrative practice during my leave of absence while undergoing cancer treatments. These visions found their way to an idea book where I jotted notes down about what the perfect practice would be like. As I brainstormed, I came up with ideas of nutrition, supplements, finding ways to work with the immune system, and not circumventing it. I worked with naturopaths and nutritionists and several chiropractors to shape my recommendations as it applied to a more holistic approach to treating illness. I have not thrown out my knowledge of western medicine; quite the contrary. I have found a way to integrate it as needed when the body's defenses break down, to help the child get well. But so many illnesses, as I said before, can be improved, or even cured, through choosing the right nutrients and staying out of the body's way of natural healing.

So I walked away from my pediatric practice of 10 years to follow God's plan, and I have never looked back. With every neck and back surgery, I continued to return to what I knew in my old practice, but God was leading me all the while to his ultimate purpose for me. My colleagues thought I was crazy to open up my own practice, especially because of my health problems. But God was telling me something quite different, that through Him, all things are possible.

Just a few short months after I completed my cancer treatment, I opened Healthy Kids Pediatrics, Center for Health and Wellbeing. As I signed the contract on my building lease, I had no fear that my repeat PET scan follow up from my cancer later that same afternoon would be normal. I created an Integrative and Holistic Practice where I embrace concepts of good nutrition, balance, and work toward improving the body's natural ability to heal through homeopathic and herbal treatments, relying on experts in Eastern medicine to guide me.

I have gotten scores of children off of ADHD medications, helped recover multiple autistic children, improved asthma and allergies, and taught hundreds of families how to live healthier. I consider my practice more of a ministry of healing, and I give God all the Glory for my success. I believe I was fully cured and kept on this Earth to teach this message to whoever will hear it and in that lies God's Ultimate Victory.

I came across an anonymous quote for a wall plaque that really spoke to me. It read:

Life is not a journey to the grave with the intention of arriving safely in a pretty and well preserved body, but rather to skid in broadside, totally worn out and proclaiming, "Wow, what a ride."

Deborah Z. Bain, M.D.

We are all on a journey. I like to think that when I die, I would have left a legacy of hope for those in my path to keep on striving to be as healthy as they can be. We only get one body on this earth and we cannot take it for granted, nor can we allow ourselves to be mere bystanders in our own healthcare decisions. We cannot afford to just accept a diagnosis and treatment plan by the doctor, without striving to determine the root of the diagnosis in the first place. So many of the diagnoses made in children and adults have a definite root in poor nutrition, poor lifestyle choices, so many things that we have control over. Only when you stop feeling like a victim and start working toward taking control of your health, can you start truly becoming well. You just have to keep asking the questions and take an active part in your health. At the same time, doctors need to be asking the right questions and using the right set of diagnostic tools in order to arrive at the true cause of your illness, not using protocol medicine and incomplete decision trees that only give them western medicine choices with which to treat.

***

"The lord is my rock, my fortress, and my deliverer, my God is my rock in whom I take refuge." - Psalm 18:2
"God is my refuge and strength and ever-present help in trouble." - Psalm 46:1

40

Chapter Five

There is Another Way

Some of the most rewarding patients I have taken care of have been the most challenging. In my previous practice, the majority of the day would be spent choosing what antibiotic or ADHD medication I would be prescribing. I would see half a dozen or more kids a day with ear infections, several with asthma, and a few kids for medication checks for their ADHD. We would see an occasional child with autism, and at least for me, that gave a sense that autism was not very prevalent. What I came to realize then, was that a traditional office is a hostile environment for some of those families with an autistic child trying to navigate through and find real answers to help their child.

The real rewards for me are not the cured ear infection on antibiotics, but it is the nonverbal autistic child that we have helped through biomedical and other approaches, who is able to look at me for the first time and say, "Hi Doctor Bain!" I have countless precious stories of children we have improved or even cured of their chronic ailments, from eczema, food sensitivities, to autism and ADHD through a different approach. There is another way that does not include drugs and Band-aid fixes on symptoms. It involves getting to the root of the problem, building a solid foundation of nutrition, and detoxifying.

***

Let me share some of my most complex cases:

Alli presented with an all too familiar story. She had been to multiple pediatricians who continued to give her medications for what they suspected was a chronic sinus infection. She also had chronic abdominal pain for which she had been to two gastrointestinal specialists. She had been to the hospital a total of eight times with no improvement in her symptoms, had tried multiple medications, and was no closer to the cause of her symptoms than when she had started. For her ailments Alli had been receiving antibiotics, and when those did not work she was given stronger antibiotics, then stronger antibiotics and steroids, then different antibiotics and stronger steroids.

When I met Alli, she had thick plaques and a thick orange discharge oozing out of her scalp. I would come to learn that that was from the overwhelming fungal infection that had developed on the aggressive antibiotic and steroid regimen she had received. She also had severe abdominal cramps and gas that was unresponsive to any pain medication. Her energy levels were so low that she was sleeping up to sixteen hours a day. The issues had become so severe that Alli's mother told me she would go into her twelve year old daughter's room when she was asleep and pray "just don't die tonight, I'll get you help tomorrow, but just don't die tonight". Mom sought me out in desperation, having reached a point of mistrust and frustration. She had heard about my integrative practice and our more holistic approach to the treatment of illnesses.

By obtaining the help of Dr Chalmers, DC, through applied kinesiology, we determined that she had a significant parasitic infection; her normal gut flora was almost nonexistent due to the aggressive use of antibiotics. We initiated a systematic approach to making Alli well, once and for all. First, we optimized her nutrients to help her body rebuild her immune system. Next we worked to clear the parasitic infection in her gut. Once the parasites were killed we were able to restore the normal gut flora with high potency probiotics and prebiotics. This was more difficult than normal due to the extensive damage from the antibiotics, as well as an overgrowth in Candida. Her response was amazing. Just by using the simple formula of "good stuff in and bad stuff out," we were able to restore her health completely, something that was not possible with previous approaches to her symptoms. Her energy returned to normal as she felt like the fog had lifted. She was able to return to school and to the game of softball which she loves. Alli is thirteen now, stands 5'10", and has more energy than any other person I have ever met. She feels amazing and is forever grateful that we gave her life back.

***

Taylor is a 17 year old who came to see me with severe joint pain, abdominal pain, no energy, sparse hair growth, weight loss, and recurrent infections. Her symptoms were so severe that she had to go on a "home-bound" status for school! She had a history of mononucleosis and was told that she had chronic fatigue syndrome. She was seen by multiple pediatricians and had an extensive workup of her vague symptoms that even included a bone marrow biopsy to rule out leukemia. All her labs were normal, so the parent was told that she would just have to take antidepressants to cope with her condition. The doctors were asking the wrong questions. We did one test that confirmed the diagnosis: –adrenal hormone saliva testing. We confirmed adrenal exhaustion when she had a flat line in her am and pm cortisol levels. We put her on compounded extended release cortisol and within a very short time, her chronic symptoms went away. She also saw Dr Chalmers to work on her nutrition. We were able to rebuild her immune system with directed nutrition to the very organs that needed help desperately. We also found that she had a parasite infection in her intestinal tract that was contributing to poor absorption of the nutrients in her diet, as well as virtually no good probiotic levels due to the antibiotics she was treated with. She had previously been unable to gain weight, had dull skin and had to be home bound for school, due to her symptoms. She is now a beautiful college student with vibrant skin and hair, and no further joint or abdominal pains. She comes back to see Dr Chalmers for a "tune up" when she is under stress with exams and lifestyle, and adjusts her supplements to help her stay well. Who would have thought that nutrition and cortisol supplementations could have such a profound effect on helping someone get well. All her symptoms were a direct result of the accumulated toxins, poor nutrition, and poor ability of her adrenals to do their job in keeping her immune system running.

Adrenal insufficiency is a likely factor in several medical conditions including hypotension, fibromyalgia, hypothyroidism, chronic fatigue syndrome, arthritis, and more. Symptoms include fatigue, insomnia, weight gain, impaired digestion, metabolism, and mental function, slowed healing, depression, hair loss, poor immune function, and cold intolerance, among others. The adrenal gland is a delicate organ that sits above the kidney. It's fundamental job is to rally all your body's resources into a "fight or flight" mode by increasing production of adrenaline and cortisol. Our bodies were not meant to live in a chronic stress state, thus creating adrenal burn-out.

***

A similar story involves Julia. She is a petite, 45 pounds, eight year old with a long history of nasal congestion, chest congestion, and stomach aches. She had a diagnosis of recurrent chronic sinusitis, reactive airway disease, dysfunctional eustacean tubes, allergic rhinitis, esophageal reflux, chronic abdominal pain, chronic adenoid and tonsilar hypertrophy, and recurrent bronchitis. She had at least eighteen courses of antibiotics, three CT scans, two xrays of sinuses, one brain MRI, and two Chest Xrays in six years. She had a negative workup for Celiac disease and Cystic Fibrosis, had normal immunoglobulins and thyroid tests, and had only a few respiratory allergies. She had also had a bronchoscopy, flex scope, adenoidectomy, tonsillectomy, sinus windows with cilia biopsy, and endoscopy and had seen a pulmonologist, two ENT's, Allergist, and GI doctor prior to her appointment. She was being treated with Pulmicort, Xopenex, Nexium, Advair, Singulair, Veramist, and Astepro. Mom reports that Julia had chronic coughs and ear infections from infancy through the age of two years old. She was treated with multiple antibiotics. When she reached kindergarten, she had constant "gunk," and was again on recurrent courses of antibiotics, mostly Augmentin. At one point, she became very pale, nauseated, and was diagnosed with functional dyspepsia and was placed on an acid blocker by the GI specialist. In the winter of second grade, she had chronic chest congestion which is when she had the aggressive workup including CT scans, T&A, sinus windows. She was better for two weeks, then developed stomach pain, nausea, and pale skin all over again. Her pediatrician thought that she must just be depressed because of all the chronic complaints, and recommended antidepressant medication!

It was at this point that Mom sought us out in desperation. She had heard of Dr Chalmers from a previous patient. He found significant kidney, adrenal, and gallbladder issues. She also had parasites and a nonexistent bacterial flora in her gastrointestinal tract. So Dr Chalmers and I went to work on our most challenging case yet. Not only did Julia have some very significant health problems to address, but she also had a mom who was extremely distrustful of the medical system.

We diagnosed Julia with chronic fatigue, bacterial dysbiosis, nutrient deficiency, and adrenal insufficiency. We worked closely to rebuild her immune system, starting with the "good stuff in and bad stuff out model." We weaned her off a few medications right away such as Singulair, Veramist, and Nexium. We then ordered IgG food sensitivities and a comprehensive nutritional panel through Genova laboratory, (NutrEval). Although she had a few sensitivities to foods, that was not her main problem. The definitive test was an am and pm Cortisol level which showed severe adrenal insufficiency.

It was only after we asked the right questions and had an open mind to search for the root of the problem, that we were able to cure Julia. She was started on Cortisol for several months and had an excellent response. Her joint pain, bone pain, abdominal pain were cured with pulling her out of adrenal crisis with the help of cortisol, along with repairing her immune system with the help of directed micronutrients, probiotics, and a healthy diet. Her neck and ear pain persisted a little longer until she allowed her neck to be adjusted, then the rest of her pain completely resolved. She is a feisty, energetic, petite, beautiful little girl who will never forget the gift of health that was given back to her.

***

Hannah is a three yr old who came to me with a history of recurrent ear infections, sinus infections, moderate to severe eczema, and recurrent abdominal pain, who had been on endless courses of antibiotics, steroids, and eventually had ear tubes. She had been on several topical steroid creams to treat the eczema without improvement. We started with food sensitivity testing and found she was severely allergic to eggs, dairy, and wheat. Further tests would confirm that she had celiac disease! She went on an elimination diet and within three months she had no signs of eczema, has not had an ear infection or sinus infection for the last year, and her abdominal pain completely resolved. Mom subsequently was tested and also had similar food sensitivities and was able to lose 15 pounds by just eliminating the foods

45

she was allergic to. She also had improvement in her nonspecific symptoms of abdominal bloating, low energy and eczema.

***

Even during my own long journey to get well, keeping an open mind and utilizing many different modalities including nutrition, supplements, and detoxification, I came across a haunting fact. I had at least two different diagnostic tests that confirmed an IgE allergy to intestinal yeast. From all my research, I was bothered by the strong link to cancers, so I sought out a more holistic minded allergy and environmental doctor to do further testing. Sure enough, he found, through standard skin prick allergy testing, that I indeed had IgE immediate allergy reactions to four different intestinal yeasts. It was the scariest experience to actually have inducible "mental fog," muscle fasciculations (twitching), severe headache, chest pain, anxiety, and abdominal pain, all due to the yeast being injected. The allergist would then reverse the reaction by a neutralizing injection. By the time I left his office, he had reversed all of the induced symptoms. He prescribed allergy shots to then "desensitize" me to these intestinal yeasts, and put me on a several month course of Nystatin. Well, not really thinking the logic through on this one, I went ahead and started allergy shots three times per week. The first few shots gave me immediate, transient mental fog, but for the next two weeks, I really did not have many detectable side effects, so as instructed, I increased the dose slightly. The theory here was to increase the antigen (yeast) injected to induce a tolerance, thereby raising the bar to which I would actually have symptoms from the presence of the yeast. Well, with the increased dose I began having a tight feeling in my right chest wall and right arm pain where most of my injections were. I started hurting up into the right side of my neck and developed fluid in my right ear. Then I started putting things together. My lymphatic system had been disrupted from the breast cancer surgery, so every bit of the yeast I was injecting into my arm was just building up and not being able to escape.

I again had Dr Chalmers help to sort things out. He tested my arm and of course, found toxins built up in the whole area. He tested my liver, gallbladder, and kidneys - all were suffering from toxic overload.

I felt like I was detoxing from chemotherapy all over again. I had a mudding procedure to my arm, castor oil packs to my chest, infrared light therapy, and handfuls of supplements over the course of 3 months to undo what I thought was a warranted series of "allergy shots." Later when my head was clear enough to think this through in my usual fashion, I thought to myself, why on earth would I want to raise the bar to which I respond to yeast in my intestinal tract in the first place. If it causes this many symptoms in small amounts, what would happen if I let it run amuck at higher levels before I even recognized it was there? And just because I have an "allergy" to yeast, does not mean I currently have an active infection. So give me Nystatin and probiotics, but I will have to say "no thank you" to allergy shots.

What do these stories have in common? There is another way. If the doctor only has a limited set of tools in his toolbox to evaluate these kids, he will arrive at the same predictable conclusions - ear infections, sinus infections, gastroenteritis, irritable bowel disease, and use the same predictable treatments - antibiotics, and "Band-aid fixes" at best. If we are to help the new generation of diseases, we first have to recognize them as real - adrenal exhaustion or insufficiency, candidiasis, bacterial dysbiosis of the intestinal tract, and use alternative treatment protocols that involve rebalancing the good intestinal flora, using nutritional supplements to support the immune system and help our body's inherent ability to heal itself as it was perfectly designed to do.

***

But He said to me, "My grace is sufficient for you, for my power is made perfect in weakness." Therefore I will boast all the more gladly about my weaknesses, so that Christ's power may rest on me. That is why, for Christ's sake, I delight in weaknesses, in insults, in hardships, in persecutions, in difficulties. For when I am weak, then I am strong. - 2 Corinthians 12:9-10
"But we have this treasure in jars of clay to show that this all-surpassing power is from God and not from us. We are hard pressed on every side, but not crushed; perplexed, but not in despair; persecuted, but not abandoned; struck down, but not destroyed." - 2 Corinthians 4:7-9

Chapter Six

My Tool Box

I have alluded many times to my "expanded tool box," so here it is. I have been adding more things to it as I continue to expand my knowledge as an integrative practitioner. As physicians we dedicate our life to learning and healing people, not as a job, but as a way of life. If we stop learning and start following a protocol or medical decision-making algorithm that is incomplete, how are we to make anyone truly well?

I have organized my tool box into four main sections which include nutrition and supplements, homeopathic or herbal remedies, more inclusive and expanded diagnostic tests, and a network of alternative medicine providers and their services. It is with these tools that I approach each patient.

A firm foundation in whole food nutrition, full of fruits, vegetables, lean meats, and unprocessed or minimally processed foods provides the body with key nutrients for normal function. I share a "Traffic Light Handout" at all my checkups which categorizes the foods we eat into green light, or grow foods, yellow light, or slow down, eat with caution foods, and red light, or stop, and avoid these foods. It is amazing how these little kids really pick up on this concept. When supplements are needed, our office provides advice and direction. Our staff is knowledgeable on the supplements we recommend, and we have a nutritionist on staff to help with more difficult cases.

Homeopathic and herbal remedies provide a very safe alternative to medications. We have a very good relationship with companies such as Nutriwest, Boiron and King Bio as well as local health food stores to provide our patients expertise on these type of supplements. Someone with a tummy ache may take a supplement with ginger, chamomile, or aloe, or a homeopathic tummy ache formula. Someone with a cold may take Elderberry, Echinacea, or a homeopathic multi-symptom cold remedy.

The main diagnostic tests I use include IgG food sensitivity testing, comprehensive digestive stool analysis, and nutritional analysis through blood and urine samples. Other tests I use include heavy metal hair analysis, celiac profile, urinary peptides for casein and gluten sensitivity, and saliva hormone testing.

Much controversy surrounds the meaning or validity of IgG food sensitivity results. I have spoken to allergists who just do not believe in them and feel it is just a reflection of the foods you commonly eat. However, when I have a child on an unrestricted diet that has serious eczema or abdominal pain, I often see only a few or even negative IgE, immediate sensitivity test results, but I have many IgG positive allergens identified on the expanded test. If the patient only eliminates the immediate sensitivity foods, they have missed more than half of the allergens causing their chronic condition. When they eliminate all of them, the eczema, abdominal pain, or whatever symptom they have, often completely goes away. Certainly this is only a small part of a workup for such chronic conditions, and we would address all aspects such as diet, food elimination, supplementation to repair deficiencies, and detoxification to achieve optimal results.

I often choose the NutrEval by Genova or a Comprehensive Wellness Panel or Organic Acid Test (OATS test) by Great Plains to determine acquired metabolic disorders, imbalances, nutrient deficiencies, and toxic elements. These have been instrumental in getting to the root of so many medical illnesses that my patients present with. It is with this information that we can formulate a nutrition plan and provide advice on supplements to help repair deficiencies.

The comprehensive digestive stool analysis is a very important test which gives a quantitative report of beneficial intestinal flora (probiotics), imbalanced flora, and dysbiotic flora (causing disease). It also identifies yeast and parasites that may be present causing disease. I cannot stress enough the importance of getting the gut happy, as 80% of the immune system is located there. I have seen so many cases of unhealthy kids that became healthy as soon as we treated the parasites, yeast, or abnormal intestinal bacteria that was present. This test is so much more sensitive than a routine stool culture, as the normal bacteria is reported as "normal flora, and is not further identified and quantified." The amount of beneficial bacteria in the intestinal tract is important also, because it gives me a place to start when it comes to recommending what they need to rebalance the gut and make it healthy. We use specific antibiotics, antifungal, and antiparasitic drugs or natural agents such as caprylic acid, citric seed extract, uva ursi, and oregano, depending on the sensitivities given on the results. I am amazed that gastrointestinal specialists as a whole do not embrace this very important tool in their profession. They rely more on invasive, costly procedures such as colonoscopy and biopsy, only to diagnose nonspecific inflammation. If they used a comprehensive digestive stool analysis as a diagnostic tool, they may yield better results and actually go farther in making a patient well, not just managed on medications for diseases such as irritable bowel disease. What if they actually had an underlying imbalance that could be cured by rebalancing their normal flora?

***

Gavin is a three yr old with Autism Spectrum Disorder, behavioral problems, and speech delays. He presented to my office for the first time with a history of persistent and recurrent ear infections for five months, for which he had pressure equalizing tubes placed, recurrent coughs, lingering for one month at a time, and recurrent lingering infections. He had a history of both expressive and receptive speech delay. His first word was uttered at twenty months and he had no two-word combinations at two yrs old. He received early childhood intervention services starting at eighteen months and then more in depth therapy from two years until present. He was evaluated by a developmental pediatrician for possible autistic tendencies.

We first did allergy testing which showed over forty foods to which he had a low level reaction including wheat and dairy. For him, it was impossible to eliminate all the reactive foods, so we recommended he start Omega 3 essential fatty acids, a probiotic, and digestive enzymes with each meal on a regular basis. Mom saw significant improvements in his behavior and speech, just eliminating casein and gluten containing products from his diet. Next we focused on evaluating him for nutrient deficiencies. His nutritional evaluation showed significant folic acid, B12, copper, iron, magnesium, and amino acid deficiencies. He also had evidence of mitochondrial dysfunction, bacterial dysbiosis, and malabsorption. Basically, he was not breaking down his food or absorbing it due to his "leaky gut" and food sensitivities, so he was becoming more and more deficient in key nutrients that in turn, affected his behavior and speech. Reversing his nutrient deficiencies and eliminating key food sensitivities has lead to dramatic improvements in Gavin's behavior and developmental delays. At a recent follow up appointment, Mom reported that Gavin has gone from the diagnosis of "autism" to just "speech delay, mostly due to articulation" when he was reevaluated by the developmental pediatrician. Even pictures from the previous year showed Gavin to be distant, glassy eyed, and withdrawn. He is now alert, observant, and interactive with people and things in his environment. He is also definitely catching up in his developmental and social skills now. He is now expressing emotion, spontaneously initiates speech, and was the most talkative person at his three year birthday party at the fire station. He is still receiving services for speech, but we are very hopeful for a full "recovery."

***

Lainey was an exclusively breastfed non-daycare child who contracted RSV at seven weeks old. She was hospitalized for six days and given aggressive treatments to improve her lung function. At four months old, she had her first "flare-up" of wheezing that is common after RSV infections. She was treated with breathing treatments of Xopenex and Pulmicort. Shortly after this episode, she started catching random viruses, but seemed to run a more severe course than others with high fevers and prolonged symptoms. It seemed that no matter what the infection, Roseola, gastroenteritis, or upper respiratory viruses, they would always trigger her wheezing, and she would again go on breathing treatments. At twelve months old, she was hospitalized again for RSV and pneumonia. Due to all her severe respiratory illnesses in the first year, she was screened by her pediatrician for Cystic Fibrosis, which was negative. She also began having fitful sleep, thrashing around, and yelling and crying inconsolably. The doctors could not figure this out, but said she would outgrow it. Mom also reported cyclical random, unexplained fevers where Lainey would have them every few weeks, lasting for several days. Her lung infections worsened throughout her second year of life, with an increase in viral and bacterial infections. She was treated with sometimes two courses of oral antibiotics. She had several trips to the ER for IV fluids, IV antibiotics and steroids At one point, she developed a severe allergic reaction to penicillin and developed an erythema multiforme rash, again sending her to the ER. Her case was referred to a pulmonologist by their pediatrician, since no matter what kind of illness Lainey had, it would end up in her lungs within days and required aggressive treatments to maintain her.

Despite strict isolation, diet changes, extreme hand washing, and major lifestyle changes - including no traveling, no play dates, no visitors, and no going out to restaurants, Lainey continued to get very sick. As she got a little older, the parents could "manage" her symptoms at home better, keeping close contact with the pulmonologist. Hospitalizations decreased, but they would have to follow up with the pulmonologist often several times per month. The specialist office suggested she be maintained on long term antibiotics and daily preventative breathing treatments including inhaled Xopenex and Pulmicort, (an inhaled steroid). Mom had the pulmonologist office run immune testing which showed only slight IgA and IgG subclass 4 deficiencies. That was it. Mom was beginning to feel very frustrated that her daughter was so sick and no one could figure out why. She decided to take a more direct role in her child's health. She went to a naturopathic physician who planted the first seed toward making Lainey well. Under his direction, mom started probiotics, vitamins, and (a fruit and vegetable supplement.) Mom was encouraged by this approach to "rebuild" her immune system, but was soon disheartened when Lainey, along with her brother, contracted the H1N1 flu virus in April 2009. It took Lainey five weeks, three trips to the ER and urgent care, and many trips to the pulmonologist to recover, while her brother was better in a few days. After she recovered, mom went back to what was recommended previously, because she was so fearful for her daughter's health. She did not know whether to put her trust in a system focused on nutrients that was unfamiliar to her, or a system of aggressive medication treatments that had a track record of bringing her through her illnesses. Mom felt defeated as she agreed to long term Zithromax antibiotics that were given three times a week, and increased when she became ill with a respiratory illness, as well as long term Pulmicort and Xopenex breathing treatments, but she continued to search for answers and to the key to make Lainey well once and for all.

However, in the summer of 2009, Lainey's condition worsened again. She started losing weight, refusing to eat, had no energy, and was shaking from being cold all the time. She had developed dark circles under her eyes and was very pale and had more fitful sleep. Mom had been pursuing ideas about the root cause of her illnesses when she stumbled on iron and its effects on the immune system. She asked the pulmonologist to run an iron level. Low and behold her level was nine, critically low. She was started on Iron therapy, as well as Vitamin D (level not tested). As her doctors continued to paint a grim picture of Lainey's future and continued to recommend strict isolation, mom started seeing some interesting effects of her natural therapy. Her daughter started sleeping through the night without fits, and having restful sleep just days after starting iron therapy. She also noticed that she started gaining weight again, had improved appetite and more energy. The family moved to Prosper into the "country" to provide a "bigger bubble" for Lainey to play and be outside.

Mom brought Lainey to our office at the age of three years, desperate for a different, more integrative, holistic plan for treating her daughter. Mom had suspected for a long time that there was an underlying cause of her daughter's medical problems, but instead of answers, she got only more Band-aid treatments for her symptoms. As you saw, mom tried Lainey on some natural supplements such as iron and vitamin D, and other vitamin and fruit/vegetable supplements that she had to pursue herself, without guidance, and was able to see improvements. But it is hard to navigate the world of supplements and natural approaches, and to weed out what is helpful and what is not. Mom sought out our help to evaluate her immune system problems in a more holistic way and provide real answers to make her well.

We started with a nutritional evaluation which showed severe B12, folic acid, Vitamin C, Zinc, and Vitamin D deficiencies. It also showed impaired methylation, critically low glutathione levels, protein malabsorption, maldigestion, and intestinal dysbiosis. She had evidence of yeast overgrowth and mitochondrial dysfunction. In other words, her immune system was a wreck.

We started her on vitamin and mineral supplements to address all her deficiencies, as well as probiotics, digestive enzymes, essential fatty acids, and amino acids. We are providing education about immune boosters, herbal and homeopathic remedies so mom can feel more confident about treating her daughter's symptoms more naturally which has been her desire all along. We expect a full and complete recovery for Lainey, not because we are using super drugs or specialized aggressive procedures to make her well. We are just giving her body what it needs to heal and minimizing the continued bombardment of drugs on her body. Most of all, we have given this family, and specifically, this child, hope that she does not need to go through life dependent upon drugs to "manage" her symptoms.

This story brings up an important point of going back to what you know and not taking a full leap of faith in something that is foreign. I know I wasted many years thinking this way through all my surgeries. But I am also here to say that there is another way and that you just have to have faith and not give up and run back to what you know when things are tough.

The final element in my tool box is the network of alternative medicine practitioners to help with my difficult cases, when I need the help of services outside the scope of my expertise. I have many chiropractors, environmental doctors, acupuncturists, naturopaths, and other healers that use different techniques such as electrodermal screening, applied kinesiology, acupuncture, and spinal adjustments to help my patients to get optimally well. For instance, I have used chiropractors to provide a safe alternative treatment for gastroesophageal reflux, plagiocephaly (acquired head flattening), constipation, headaches, and numerous other ailments. I have used alternative practitioners' skills in applied kinesiology (muscle testing) and electrodermal screening as a helpful diagnostic tool for those patients with an allusive diagnosis that cannot be easily determined by conventional tests. These alternative diagnostic tools focus more on imbalances in the body and offer help to restore normal function through homeopathics or micronutrients. Sometimes there just isn't a "medical diagnosis" to be made, but rather a "metabolic disruption or imbalance" that needs to be addressed in order for the body to function properly. If these warning signs of the body go unchecked, then the natural evolution of the imbalance is to develop into a "disease state," for which western medicine has a drug to treat. The key is to balance the body and not ignore the warning signs, so we do not go down the path of disease.

Albert Einstein once said, "insanity is doing the same thing, but expecting different results." We are creatures of habit. In turn, we have identifiable behavior patterns that we use each day. Likewise, as physicians, we tend to use common tools. We use the same antibiotics, order the same tests, refer to the same specialists, and recommend the same approach to treat a medical problem. The best way to get people to change their behavior is to change the systems that affect them. Great systems don't have to be complicated, but they do have to produce measurable, reproducible, and positive results. I can honestly say that if more physicians added just a few of these new tools to their diagnostic toolbox, they would be taking a step toward becoming a great doctor, not just a good one, and better yet, their patients would be much closer to becoming truly well.

***

"The mocker seeks wisdom and finds none, but knowledge comes easily to the discerning" - Proverbs 14:6

Chapter Seven

A Simple Solution

I have presented so many of my complicated cases throughout this book to hopefully show a pattern that is predictable and reversible, but it is also my intent to equip my readers with the knowledge of how to treat the most common complaints naturally before they become a larger problem. This chapter is dedicated to some of those simple solutions.

In my opinion, the skin is the window to the immune system. So when I see kids with eczema, I know their immune system needs help. In the most simplistic sense, there are two main types of eczema, the rashy ones and the dry scaly ones. Rashy eczema usually means there is some food or environmental allergy flaring it. In my experience, the majority of rashy eczema in babies is caused by food allergies. In babies, the most common food allergy is the milk-based formula or something in mom's breast milk the baby is reacting poorly to. The top reactive foods include cow's milk, soy, egg, wheat, peanuts, and tomatoes. Incidentally, these are the same foods we tell moms to eliminate from their diet for babies with colic. We encourage moms to change to a hypoallergenic formula such as Alimentum, Nutramagen, Elecare or Neocate or to eliminate these foods from mom's diet for a few week period, one at a time, to determine what may be causing the symptoms. Once the offending food has been identified and the trunkal rash seems to be improved, then the unoffending foods can be reintroduced also one at a time.

***

Let me share an example of a patient to illustrate this point.

Jaden was a 4 month old at his first visit to my office. He presented with severe, oozing eczema with a crusty, weepy, cracking scalp with severe erythema and signs of a staph impetigo infection. He had cradle cap shortly after birth that was just dry and crusty, but then after the first set of vaccines with Hepatitis B, Hib, Prevnar, IPV, Rotavirus, and DTaP, he developed severe head to toe eczema. He was initially on Similac Advance formula which we changed to soy in this Asian child. He failed to improve. We obtained an Immunocap IgE childhood allergy panel which showed a soy allergy! We changed him to Nutramagen which failed to improve his eczema, so we tried Neocate which was exactly what he needed. Neocate is a very hypoallergenic amino acid formula. He did well on this until mom started offering him infant cereal. He had a brief flare of eczema on his arm creases, but quickly improved after removal of the food from his diet. He had recurrent bouts of impetigo from the affected areas and was treated with Bactroban ointment and Zinc oxide diaper cream. He weaned off Neocate at 15 months to Goat's milk because he still could not tolerate cow's milk. He has done remarkably well in the last year and his eczema has completely cleared.

It is so rewarding to me to transform this irritable, continually scratching, poor sleeper with severe eczema into this energetic, healthy, child with beautiful skin. During this time we kept his vaccines to a minimum, resuming them only after he was on the right formula and his eczema had resolved. We also treated him with probiotics to restore normal intestinal flora to his digestive tract and started a supplement of essential fatty acids in the form of flax oil daily. Products such as coconut oil, olive oil, and calendula cream, are very effective moisturizers and are also more nourishing to the skin than the petroleum products, and far safer too.

You can no more consider the skin in isolation of the rest of the body when it comes to treatment, just as I could not do the same for my neck and back. The compartmental model does not work here either. You cannot just apply steroid creams and treat with antihistamines for it to go away. Remember, eczema is coming from a problem from within. Often the eczema is related to multiple systems including the gastrointestinal and endocrine systems, and will not get better until you treat it accordingly.

The dry, scaly eczema is related to an Omega 3 essential fatty acid (EFA) deficiency and rarely, zinc deficiencies. This can simply be remedied by adding one teaspoon of flax oil to the child's diet and one tablespoon to mom's diet while breastfeeding. Sometimes with severe eczema there is a poor enzyme conversion of LA (linoleic acid) to GLA (gamma linolenic acid). Studies have shown that rubbing 2 capsules of Evening Primrose oil (major source of GLA) on the skin two times a day circumvents this enzyme step and provides more of this anti-inflammatory Omega 6 EFA. We will often encourage probiotic use daily for these kids because as I said, it is often related to the unhealthy digestive tract.

The longer eczema goes untreated in a more holistic way, the more likely the effects will become more far reaching. If eczema is food allergy related, and the offending food is not recognized, a leaky gut situation will develop, leading to more food sensitivities and finally to nutrient malabsorption which has its own set of consequences. The whole cycle is perpetuated by the Band-aid approach to the treatment plan, giving parents the false sense that the eczema is gone and that everything is fine. Sure, some kids outgrow some transient eczema such as when they wean off formula and refuse to drink milk as toddlers, but some are much more severely affected and develop numerous allergies and nutrient deficiencies which may eventually lead to behavioral problems, speech delays, and even autism.

Upper Respiratory Infections (URI) or common colds are a common in children. However, since the cold medications were taken off the market for these young kids, parents are often left to scramble to provide symptomatic relief of their child's symptoms, and often do not know where to turn. At Healthy Kids Pediatrics, we talk about "immune boosters" as a collective term meaning those things that boost the immune system and help the child fight off the cold. These can be homeopathics or herbal remedies, or even Vitamin D and Vitamin C. Approximately 80% of the children I test Vitamin D level on, are deficient in that vitamin. Vitamin D is very involved in keeping the immune system running and in making white cells, so if a child is deficient, especially in the winter when he is not getting a lot of outside time, then it makes sense that he would succumb to many viral illnesses. A simple supplement of Vitamin D through the winter, along with an immune booster such as Elderberry or Briar Rose may significantly reduce the development of respiratory illnesses. With more and more uncertainty as to the safety of cold preparations in children, why not choose a natural product next time your child gets sick or has a fever. Companies such as Boiron, King Bio, and Hylands make safe homeopathic remedies for fever, cold symptoms, flu and even teething.

Some respiratory illnesses progress into sinusitis which is routinely treated with antibiotics. Treatment alternatives may include colloidal silver, oil of oregano, corriolus, saline rinses to nasal cavities. Often parents see green nasal discharge and think infection, but it may just be the normal progression of the cold. Viruses last up to three weeks. It is my recommendation that parents do not rush to treat with an antibiotic, but rather work with the body's own immune system to help it fight off the illness. Chiropractors have been very helpful in relieving sinus pressure and helping with lymphatic drainage by massage over the frontal and maxillary sinus areas.

That brings me to ear infections, one of the most common diagnoses made by pediatricians and one that antibiotics are over-prescribed. Symptoms of an ear infection include earache, fullness and pressure in the ear, and fever. They may be irritable or restless, have a nasal discharge, diminished appetite, or cry at night when lying down. Common complications may include conductive hearing loss, perforated ear drums, and chronic ear infection or "glue ear." Less common complications include speech delay and even mastoiditis where the infection extends into the mastoid bone of the skull.

Acute ear infections can be caused by an upper respiratory infection that spreads to the ears. Inflammation from the infection causes the Eustachian tube at the back of the throat to swell shut, trapping bacteria in the middle ear cavity. The Eustachian tube provides an outlet for mucus and equalizes pressure changes. Normally closed, it opens to allow drainage of mucous into the throat. If this does not work properly, an air pocket forms and the negative pressure in the middle ear space then pulls the bacteria and/or viruses into the middle ear where they can flourish.

Specialized cells in the middle ear manufacture fluid that helps keep out invading germs. If the Eustachian tube becomes so swollen that the fluid becomes trapped in the middle ear, the area can become inflamed. If infection sets in, the fluid in the middle ear cannot escape. The earache is usually due to inflammation behind the eardrum, often from a plugged up Eustachian tube. A child's Eustachian tube is shorter and lies in a more horizontal position than adults, thus making children more prone to ear infections.

While bacteria can be the direct cause, an ear infection is usually the result of an "insult" to the body, such as an allergic reaction or a weakened immune system. Imbalances in the digestive system, irritation or inflammation of the mucous membranes, food or respiratory allergies, consumption of cow's milk, can all lead to excess mucus secretion and thereby make a child more susceptible to the ear infections. While antibiotics do kill harmful bacteria, they also indiscriminately destroy beneficial intestinal bacteria. This upsets the delicate balance in the digestive tract, leading to overgrowth of unwanted bacteria and fungus such as Candida Albicans. Unchecked, these organisms can cause further problems and create a vicious cycle of more ear infections, more antibiotics and more imbalance.

At Healthy Kids Pediatrics, we offer a better solution to treating ear infections and work to prevent their reoccurrence by focusing on strengthening the child's immune system and focusing on the underlying cause of the infection in the first place. We offer food allergy testing especially when a child has had persistent or recurrent ear infections, as well as refer to chiropractors to help with adjustments to promote drainage of the Eustachian tubes. We have also found that garlic oil, neem oil, or colloidal silver ear drops have been helpful in treating the persistent fluid behind the tympanic membrane. These provide antiviral, antibacterial and anti-inflammatory effects directly at the site of ear fluid or infection. They are sold at healthfood stores as "ear infection drops." They can also be used at the onset of a cold in a child who has a tendency to have ear infections, as one way to prevent fluid buildup in the middle ear space. They cannot be used if a child has pressure equalizing tubes in place. One to two drops of peppermint essential oil can be massaged in front of the ear and down the neck to help promote Eustachian tube drainage in conjunction with massage in school-aged children. Peppermint oil can sting, especially if it get in the eye, so make sure your child is old enough to know better.

When we do choose antibiotics to treat an acute otitis media infection, we do so with great care not to disrupt the natural balance in the intestinal tract. We recommend probiotics such as Florastor (Sacromyces boulardii), a beneficial intestinal yeast that is not killed by antibiotics, to use while on antibiotics. Once the antibiotics are completed then we recommend restarting a Lactobacillus acidophilus product. We have also recommended colloidal silver orally as a natural alternative for treating ear infections, as it has antiviral and antibacterial properties. Occasionally we will have a child with persistent "glue ears" and conductive hearing loss who has not responded to natural treatments. Usually they have a family history of ear infections and tubes in one of the parents. Instead of continuing antibiotics we would recommend pressure equalizing tubes at that point, but this is definitely the exception at our office.

While this is just a short list of common illness we treat more naturally, there are many books on natural baby care that I encourage my parents to consult on specific ailments. Kids have a great capacity to heal themselves if we stop getting in the way by over-treating them with pharmaceuticals.

***

Do not conform any longer to the pattern of this world, but be transformed by the renewing of your mind. – Romans 12:2
Be confident in this, that he who began a good work in you will carry it on to completion until the day of Christ Jesus. – Phillipians 1:6

Chapter Eight

Our Future

I have these final thoughts to share. If you make just one positive choice toward health today, you will be healthier today than you were yesterday. If you make a pattern of good health choices, just think about the possibilities six weeks, or even six months from now. We are all on a journey in this life. It is much easier to go through life optimally well and enjoy life to the fullest. Fighting lengthy illnesses just steels your life away. I know it did for me. I spent my entire 30's in doctor's offices and either having surgery or recovering from it, just to end in a battle with breast cancer. I can't have those years back, but I can look toward the future with hope and confidence, knowing I have invested in my health for years to come. I can truly say that I am healthier now than ever, but it took hard work and dedication to get here. But you know what, I am worth it, and so are you!

I see many children in my office that are a wreck medically and developmentally, and they have a laundry list of medications that doctors have put them on. It breaks my heart seeing families going through chronic illnesses with their kids. However, I hope that you take away from the case histories I presented, that these kids did not get there over night. A chronic illness develops when something breaks down in their immune system to cause the symptoms. These warning signs, however, are often are ignored or not recognized. I hope through this book families will recognize early, the patterns or associations I have laid out so they can prevent some of these predictable diseases. It is up to us as physicians to keep asking questions, not stop at the easy fix, and to help these families raise healthy kids, not medically managed kids. We owe it to them. It is up to the parents to not give up, to be educated, and to embrace a healthier approach to raising their kids, not just take a back seat approach to their health.

We cannot continue to believe that the toxins, drugs, vaccines, poor food choices, and poor food quality have nothing to do with the evolution of a whole generation of chronically ill children. As physicians, we should always be searching for the ultimate, underlying cause of illness, not just stopping at the first few symptoms because we can give them a drug to make them feel better. I have outlined multiple diagnostic tools that I have used to help me identify the underlying cause of a child's symptoms. Most of them pertain to evaluating the metabolic disruptors causing disease and the deficiencies in essential nutrients the body needs to be healthy, such as vitamin, mineral and essential fatty acid levels. There is a place for more specialized testing, but they should not be done in exclusion of the building blocks identified here. I do not routinely recommend expensive and invasive tests, but rather to use an open mind and listening ears to help people get to the root of their problem and help them become well. Using my simple approach of "good stuff in" and "bad stuff out" has helped me accomplish this. I believe that if we identify these things early on, we can prevent or even reverse the development of chronic disease.

When did we throw out common sense for protocol medicine. These protocols were meant to be a guideline for physicians, not something to hide behind when someone has a poor outcome such as becoming autistic after a vaccination that was on the "immunization schedule." I have seen more and more protocols being created such as giving the Hepatitis B vaccine in the newborn nursery since I started practicing medicine fifteen years ago. Some protocols are lifesaving such as CPR, and need to be followed without exception, but even these guidelines go through revision every few years as they find what works better. The only revision to the vaccine protocol is to add more vaccines to the schedule, not to make them safer. If we are to have protocols for vaccines, antibiotics, Vitamin K administration in newborns, then they should be based on real risk of disease and revised often, not perpetuated without exception. What about adding oral Vitamin K protocols as an option in the newborn nursery, or offering probiotics routinely for any antibiotic administration whether IV or oral?

Our kids are our future. How can we continue to ignore the sheer toxic load our children are under and not do anything about it? How can parents continue to have blind faith in a system that is so out of touch with what "healthy" really means? Health is not just the absence of a disease. It is also not just being well managed on medication either. There is another way, and hopefully you have been inspired to find yours after reading this book.

I do not usually get involved in politics, but I can no longer be silent on this issue. This is not the time for government run healthcare system, but for the best integrative healthcare system in the world. Even with the current problems with our healthcare delivery system, it is far better than a plan without choice. In order to have a government for the people and by the people, the people have to show up! No matter where you stand on issues, make your choice heard, or some out of touch politician will make policy decisions that are starkly opposite of your beliefs. Right now, politicians are pushing to ban supplements as we know them. Supplements would be made illegal to purchase without a doctor's order. That is a very scary thought to me, considering most physicians do not have a clue on vitamin supplements. It took me years to learn how to incorporate them into my integrative practice. I am also concerned about politicians pushing for mandated vaccinations in Texas, my home state. I do not know about you, but I do not think politicians have any right to make policy for something they do not understand, and have no medical expertise in. In a government run medical system, doctors would be mere robots to carry out systems based on protocols that do not include any alternative medicine options.

It is up to us to stand up for our rights. Demand excellence not mediocrity from your doctor, or change doctors. If you always get a drug when you visit the doctor, and don't even know why, or worse yet, don't feel any better after taking it, then it might be time to change your approach to the problem. Always remember the "good stuff in and bad stuff out model" to help you. You need to be a participant in your own healthcare, not just a bystander.

My plea to parents is that they become advocates for their own children's health. Our kids are our legacy. Ask the hard questions. Do your own research. Go to a pediatrician that will work with you and guide you in how to raise a healthy child.

My plea for health care providers is to "first do no harm." Remember your Hippocratic Oath. Listen to parents concerns. Be open-minded. Take better medical histories to specifically identify autoimmune disease and mental illness. Some kids are just not healthy enough to get the full set of vaccinations. Work with families to develop a custom vaccination schedule that works, but be willing to reassess if the child is not doing well with them. Do not penalize or ridicule families for doing a modified or selective schedule.

My plea to researchers is to make cleaner vaccinations. Remove the aluminum and other dangerous substances from the vaccinations. Find ways to make vaccinations work with the body's natural defenses, not circumvent them. The public mistrust in vaccines is well warranted when recalls of vaccine for contamination have become a common occurance.

My plea to the states and schools is that vaccinations not be mandated for school entry. At some point we lost sight of the good of the child and traded it for "the common good." But as Autism has risen to astronomical heights and has now reached 1:91 kids, I think doing a selective or modified vaccine schedule has now become "the common good," and is just the right thing to do.

Lastly, my plea to women thinking of becoming pregnant is to undergo a detoxification program first, eat healthy foods and taking supplements to improve your health before bringing another child into the world.

***

Jesus said, "It is finished." - John 30
"Well done my good and faithful servant." - Matthew 25:23

Deborah Z. Bain, M.D.

A portion of the proceeds from the sale of this
book will be donated to autism research.

NOTES

NOTES

NOTES

Deborah Z. Bain, M.D.

NOTES

NOTES